Electric Vehicle Batteries

Electric Vehicle Batteries

From Sourcing to Second Life and Recycling

Edited by

BOB GALYEN
Galyen Energy LLC, Noblesville, IN, USA

FRANK MENCHACA
Auzolan LLC, Pittsburg, PA, USA

WILEY

Copyright © 2025 by John Wiley & Sons Inc. All rights reserved, including rights for text and data mining and training of artificial intelligence technologies or similar technologies.

Published by John Wiley & Sons, Inc., Hoboken, New Jersey.
Published simultaneously in Canada.

No part of this publication may be reproduced, stored in a retrieval system, or transmitted in any form or by any means, electronic, mechanical, photocopying, recording, scanning, or otherwise, except as permitted under Section 107 or 108 of the 1976 United States Copyright Act, without either the prior written permission of the Publisher, or authorization through payment of the appropriate per-copy fee to the Copyright Clearance Center, Inc., 222 Rosewood Drive, Danvers, MA 01923, (978) 750-8400, fax (978) 750-4470, or on the web at **www.copyright.com**. Requests to the Publisher for permission should be addressed to the Permissions Department, John Wiley & Sons, Inc., 111 River Street, Hoboken, NJ 07030, (201) 748-6011, fax (201) 748-6008, or online at **http://www.wiley.com/go/permission**.

The manufacturer's authorized representative according to the EU General Product Safety Regulation is Wiley-VCH GmbH, Boschstr. 12, 69469 Weinheim, Germany, e-mail: Product_Safety@wiley.com.

Trademarks: Wiley and the Wiley logo are trademarks or registered trademarks of John Wiley & Sons, Inc. and/or its affiliates in the United States and other countries and may not be used without written permission. All other trademarks are the property of their respective owners. John Wiley & Sons, Inc. is not associated with any product or vendor mentioned in this book.

Limit of Liability/Disclaimer of Warranty: While the publisher and author have used their best efforts in preparing this book, they make no representations or warranties with respect to the accuracy or completeness of the contents of this book and specifically disclaim any implied warranties of merchantability or fitness for a particular purpose. No warranty may be created or extended by sales representatives or written sales materials. The advice and strategies contained herein may not be suitable for your situation. You should consult with a professional where appropriate. Further, readers should be aware that websites listed in this work may have changed or disappeared between when this work was written and when it is read. Neither the publisher nor authors shall be liable for any loss of profit or any other commercial damages, including but not limited to special, incidental, consequential, or other damages.

For general information on our other products and services or for technical support, please contact our Customer Care Department within the United States at (800) 762-2974, outside the United States at (317) 572-3993 or fax (317) 572-4002.

Wiley also publishes its books in a variety of electronic formats. Some content that appears in print may not be available in electronic formats. For more information about Wiley products, visit our web site at **www.wiley.com**.

Library of Congress Cataloging-in-Publication Data applied for:

HardBack ISBN: 9781394262113

Cover Design: Wiley
Cover Images: © 3alexd/Getty Images, © joel-t/Getty Images, © Just_Super/Getty Images, © Agaten/Getty Images

Set in 10/12pt STIX Two Text by Straive, Chennai, India

Contents

LIST OF CONTRIBUTORS vi
FOREWORD vii
PREFACE viii

Introduction 1
Bob Galyen and Frank Menchaca

1. **High-Capacity Battery Technology** 9
   David Howell and Tien Duong
2. **Sourcing Lithium-Ion Batteries** 27
   Austin Devaney and Bob Galyen
3. **Battery Design** 53
   Joern Tinnemeyer
4. **Vehicle and Infrastructure Integration** 83
   Oliver Gross
5. **Lithium-Ion Battery Recycling** 125
   Steven Sloop
6. **Electric Vehicle Batteries: Repurposing for Second Life** 149
   Apoorva Roy, Hamidreza Movahedi and Anna Stefanopoulou
7. **Designing New Engineering Teams and Practices** 189
   John Warner

JOB LISTINGS 219
INDEX 221

List of Contributors

Austin Devaney
Li7Charged
Brevard, NC, USA

Tien Duong
Department of Energy, Vehicle Technologies Office
Washington, DC, USA

Bob Galyen
Galyen Energy LLC, Noblesville, IN, USA

Oliver Gross
Electrical Energy Technology
Stellantis, Auburn Hills, MI, USA

David Howell
Strategic Marketing Innovations (formerly with the Department of Energy)
Washington, DC, USA

Frank Menchaca
Auzolan LLC, Pittsburgh, PA, USA

Hamidreza Movahedi
Department of Mechanical Engineering
University of Michigan, Ann Arbor, MI, USA

Apoorva Roy
Department of Mechanical Engineering
University of Michigan, Ann Arbor, MI, USA

Anna Stefanopoulou
Department of Mechanical Engineering
University of Michigan, Ann Arbor, MI, USA

Joern Tinnemeyer
EnerSys
Reading, PA, USA

Steven Sloop
OnTo Technology LLC
Bend, OR, USA

John Warner
Holly, MI, USA

Foreword

This book differs dramatically from most texts on batteries, which predominantly focus on the chemistry and materials science of the active components. In contrast, here the authors focus on the engineering of batteries and the creation of a sustainable ecosystem, with a particular emphasis on electric vehicles. This book is essential reading if we are to have a sustainable planet, not one where it takes more than 40 kWh of energy to produce a 1 kWh lithium-ion battery and some materials traverse tens of thousands of miles from the mine to the finished product. Not one where we use toxic chemicals, like N-Methyl-2-Pyrrolidone (NMP)for coating the cathode material, that subsequently require the most energy to remove during the manufacturing process.

We need to relook at the whole manufacturing process, from mine to finished product. Instead we need to leapfrog today's methods to build the next generation factories that are sustainable and can use mostly recycled components. At the front end, we must use mining technologies that do not generate mountains of toxic solid waste and/or wastewater. An example might be the plans to drill for lithium as proposed in Arkansas, then separate the lithium from the other cations before pumping those other cations back into the same ground. At the backend, we have to get away from totally destroying old batteries into black mass, but rather get the materials back as far as possible in their virgin form and do it locally.

As we implement these changes, safety must always take first place. We cannot afford to have low-quality batteries enter the market, as this could damage the future market, particularly when human life is compromised. At the same time, it is crucial to remember that all forms of energy is inherently unsafe. Given today's safety standards, it is unlikely that internal combustion engines, where a 20-gal fuel tank is located beneath the back seat (occupied by a 2-year-old child), can be considered a reasonable safe technology.

This book serves as a guided tour through all of these issues and more, written by experts practicing in the fields they discuss.

Sir M. Stanley Whittingham
Binghamton, NY

Preface

The first decades of the 21st century have been a period of profound change for engineering. The rapid development of technologies such as advanced manufacturing and artificial intelligence, alongside the post-pandemic onshoring of manufacturing, converged with a shift from fossil fuels to battery electric and renewable forms of energy. These factors have necessitated, if not a reimagining, then a significant revision of how vehicles, propulsion, and infrastructure are designed and built.

This is the first in a series of publications designed to identify, define, and implement these revisions. Each book in this series focuses on an area of critical importance to new transportation: batteries, metals, composite materials, and alternative fuels. It explores what engineers, aspiring engineers, and the instructors and leaders who support them, need to know about each of these topics.

These are not research endeavors, although heavily informed by research, but rather practical publications. The series is authored by industry experts who spend each day of their working lives to figuring out how to implement these profound changes in their practices.

In addition to in-depth discussions of key strategies for successfully and safely designing vehicles and putting them into use, each book offers case studies of businesses and organizations at the forefront of this new world. It also provides information on standards and regulations and profiles the kinds of jobs and skills that are emerging. The objective of the series is to help engineers work safely and effectively as they confront the new challenges associated with technology and the energy transition.

Frank Menchaca
Pittsburgh, PA
January 2025

Introduction

Bob Galyen
Galyen Energy LLC, Noblesville, IN, USA

Frank Menchaca
Auzolan LLC, Pittsburg, PA, USA

What You Will Learn in This Chapter

This chapter introduces you to this book: why we wrote it and what sustainability has to do with engineering and batteries.
  In this chapter:

- We describe the problems the book sets out to address and how we organize our approach.
- We explore the drivers of sustainability in transportation.
- We examine how those drivers affect the development of vehicles and the engineering practices that support them.
- We include terms to know, information on jobs that relate to the topics at hand, and resources for learning more.
- We use a real-world case study to illustrate our points, as with most topics in this book.

Case Study: Circular Economy for EV Batteries in Australia

In 2021, a team of researchers at the University of Melbourne set out to study the impact of reusing electric vehicle (EV) batteries.[1] Some projections had EVs accounting for 30 % of vehicles on Australia's roads by 2035. This growth was important to decarbonatization of transportation and to achieving net zero emissions, but it would not be consequence free for the environment.

While EVs produce no tailpipe emissions, their batteries require impactful processes such as mining to reach underground reservoirs of brine and other materials. Refining brine produces lithium, a chemical element critical to power in EV batteries. This involves large quantities of pumped water, which also means producing wastewater. Besides, emissions associated with mining equipment, transportation, and manufacturing and EV batteries threaten to neutralize the gains made by the vehicles they power.

Research has already established that an EV battery, once it serves the vehicle's 8 to 10-year average lifespan, could retain as much as 80% of its capacity.

(continued)

---

[1]Nicholas Wilson. Resources, Conservation & Recycling 174 (2021) 105759.

Electric Vehicle Batteries: From Sourcing to Second Life and Recycling, First Edition.
Edited by Bob Galyen and Frank Menchaca.
© 2025 John Wiley & Sons, Inc. Published 2025 by John Wiley & Sons, Inc.

Harnessing that capacity without dismantling the battery and interacting with its potentially toxic contents, known as black mass, became the goal of the Melbourne team and to achieve it, they conducted what is referred to as a Lifecycle Assessment (LCA). An LCA is the study of a product or procedure's environmental impact – a codification of the steps that go into manufacturing something and quantification of their consequences: the energy they use, the emissions they produce, and more. LCAs constitute a critical tool in identifying where environmental tolls are heaviest and how to reduce them.

The researchers defined and assessed not only the creation of EV batteries but, just as important, their use as storage devices when their service in the EV ended, known as their second life. This meant mapping out any remanufacturing and transportation, assessing their impact, and establishing how much energy the repurposed batteries could store as offset.

This EV battery lifecycle consists of five phases in two cases of reuse, one as a home energy storage system (HESS) and one as home energy battery pack (HEBP):

1. Minerals extraction and manufacture, in which lithium and other natural materials are mined and refined, occurring in China and the Rest of the World (ROW).
2. EV use, in which the battery powers the car over its 8- to 10-year life span.
3. Repurposing, in which the battery is remanufactured so that it can be reused. Because the study assumes the battery was produced with the intention of being reused, it allocated 25% of the emissions generated in its production to reuse. In the study, as in other LCAs, energy use and emissions must always be accounted for; in this case, allocation was done over the period of initial use and reuse.
4. Use, in which the battery is put to reuse in a HESS and HEBP.
5. End of Life, in which the battery is dismantled and its constituent units are recycled.

By assessing each step in the repurposed EV battery's initial and second life, the Melbourne team was able to draw some important conclusions about the environmental benefit of recycling – or creating a circular economy – for batteries and their contents. "The repurposed battery has a smaller footprint across all eight environmental impact categories, provided it operates for a minimum of six years[2]," they wrote. The environmental categories on which the team studied the battery's impact included:

- Global warming potential (GWP)
- Terrestrial acidification potential (TAP)
- Surplus ore potential (SOP)
- Fossil resource scarcity potential (FFP)
- Water consumption (WCP)

These are important areas that directly affect human health, biodiversity, and the economy, among other things. Understanding the impact of manufacturing on these areas, and finding means of limiting that impact constitutes the essence of a sustainable engineering practice. That is the focus of this book.

---

[2]Wilson, p.

A New Age in Engineering

The Melbourne case illustrates several important transformations the field of mobility engineering is undergoing. One is a commitment to sustainability. A term used frequently and in a wide variety of contexts, sustainability can seem general and vague. Our use of the term derives from the 1987 United Nations Brundtland Commission's definition: "meeting the needs of the present without compromising the ability of future generations to meet their own needs."[3] Natural resources are finite, says this definition, and we must use them in a way that does not damage the world and its people and prevent future generations from enjoying their benefits, thriving, and leading prosperous lives.

Sustainability has become critical to mobility because transportation accounts for an estimated 20% of global greenhouse gas (GHG) emissions – second only to electricity generation.[4] These are the emissions, produced by burning fossil fuels burned in cars, airplanes, ships, and other forms of transportation. Their accumulation in the atmosphere causes global warming. Governments and businesses throughout the mobility industry have committed to transfer to electric and renewable energy sources as a means of reducing GHGs. For many, this is to comply with the Paris Agreement,[5] a 2015 international agreement to limit global warming to less than two degrees centigrade by 2050 – a threshold at which climate scientists project the impact of climate change to be manageable – by reaching net zero GHG emissions.

This energy transfer that supports this goal constitutes one of the largest – perhaps the largest – changes transportation has undergone since its beginning. Batteries play a central role. Let us examine how:

- Princeton University's Net Zero America study comprehensively calculates what will be required for the United States to achieve net zero emissions by 2050. It presents five scenarios and the backbone for all is a shift from fossil fuel-powered vehicles to EVs. The study projects a future in which electric vehicles replace internal combustion engine vehicles (ICEVs) in a vertiginous climb: 49 million in 2030, 204 million in 2040, and 328 million by 2050.[6] All of these vehicles will likely require batteries.

- In 2022, the United States passed the Inflation Reduction Act (IRA). Through a series of tax credits to businesses and individuals, the IRA is intended to attract EV buyers and accelerate battery production in the United States. The bill was accompanied by federal investments such as in 2023, when the U.S. Department of Energy loaned Ford Motor Company and Korean battery producer SK $9.5 Billion to build, among other things, a series of battery factories.[7] This enabled the automaker to make the largest financial investment in its history – in batteries.

---

[3]**https://www.un.org/en/academic-impact/sustainability**
[4]**https://www.statista.com/topics/7476/transportation-emissions-worldwide/#topicOverview**
[5]**https://unfccc.int/process-and-meetings/the-paris-agreement**
[6]Net Zero America, p. 20.
[7]**https://news.yahoo.com/ford-just-got-loan-bigger-180**000907.html#:~:text = The%20Cool%20Down-,Ford%20just%20got%20the%20biggest%20U.S.%20investment%20%27since%20the%20advent,venture%20is%20spending%20it%20on&text = The%20U.S.%20Department%20of%20Energy,to%20build%20electric%20vehicle%20factories.

- In 2023, the European Union (EU) adopted a sweeping new regulation governing the production of batteries and the tracking of battery materials. Scheduled to go into effect in 2026, the legislation addresses the entire battery lifecycle. The law states:
  - In view of the strategic importance of batteries, to provide legal certainty to all operators involved and to avoid discrimination, barriers to trade, and distortions in the market for batteries, it is necessary to set out rules on the sustainability, performance, safety, collection, recycling, and second life of batteries as well as on information about batteries for end users and economic operators. It is necessary to create a harmonized regulatory framework for dealing with the entire life cycle of batteries that are placed on the market in the Union.[8]
- By the end of the 21st century's first decade, China had established itself as the global leader in both battery and EV manufacturing. Between 2009 and 2022, the Chinese government invested over 200 billion RMB, or $29 billion USD[9] in battery and EV technologies, with one Chinese company (CATL) emerging as the largest battery manufacturer in the world.

These are just a few examples of the centrality that batteries have attained within industry and government worldwide. This ecosystem is vast and dynamic. The Melbourne case study illustrates that this transition is complex and multifaceted and requires executives, engineers, designers, and technicians to do their work differently. Designing for reuse and understanding the impact of sourcing and production through LCAs are just a few of the many practices mobility professionals must learn and utilize. This change affects nearly everything: the structure of teams, manufacturing, profitability, professional development, and more. We are truly entering a new age of sustainable battery engineering, one that is critical to the future of mobility.

The purpose of this book and the series to which it belongs is to define and describe this new age in order to help businesses, government, and people thrive while supporting the goal of net zero emission transportation. We have chosen to begin with batteries because they are sustainable mobility's most significant and nearest-term power sources.

Our approach is very application oriented: we want to provide concrete guidelines that engineers and people who want to be engineers can follow in actual practice. Consequently, we engaged industry practitioners – people who are actually doing the work of manufacturing batteries and EVs – as authors. Getting working mobility professionals to contribute to a book had its challenges; at the same time, we genuinely hope this approach constitutes its distinguishing value.

How This Book Is Organized

Over seven chapters this book will explore every aspect of the battery ecosystem.

- High Capacity Battery Technology sets the stage by focusing on Lithium-ion batteries, which are the primary power and storage sources in EVs. David Howell and Tien Duong review the chemical and structural components of these batteries and provide insights on technical advances in energy density along with

---

[8]**https://eur-lex.europa.eu/eli/reg/2023/1542/oj**
[9]Yang, Zeyi. "How did China come to dominate the work of electric casrs?" MIT Technology Revie. February 21, 2023. **https://www.technologyreview.com/2023/02/21/1068880/how-did-china-dominate-electric-cars-policy**

the trade-offs that come with each new change. In fact, trade-offs are a key topic our contributors focus on in this book. Technical change always comes with strings attached, whether those strings be in costs, power, or environmental impact. Informed choices are critical to commercial success and to sustainability and we have illuminated those choices wherever possible. This chapter provides readers with a foundation for understanding the subsequent topics.

- Sourcing addresses how the batteries introduced in the previous chapter begin their lifecycle. Austin Devaney and Bob Galyen review different types of sourcing with a specific emphasis on the environmental impacts of each. For example, lithium is sourced from brine which involves the resource-intensive process of pumping and evaporation. They guide the reader through the consequences of sourcing, highlighting opportunities for adopting sustainable material extraction practices.

- Joern Tinnemeyer's chapter on Battery Design clarifies the many trade-offs that come with design decisions. It foregrounds that discussion by differentiating between cell types and their characteristics. It examines the continual interplay between performance and cost. It also provides an understanding of battery cell composition in terms of safety, especially where it concerns thermal runaway. Designing for safety is critical to EV acceptance and this experienced professional guides the reader through the many decisions that underpin safe and effective performance.

- Oliver Gross provides a comprehensive discussion of Vehicle and Infrastructure Integration. He elucidates the challenges that go along with designing battery-powered vehicles for an infrastructure that is still emerging. Gross gives readers an understanding of the essential differences between EVs and internal combustion engine (ICE) vehicles and how they guide design decisions and affect costs and efficiency. An in-depth investigation of charging as a design also helps readers understand the new and important design considerations.

- In Steven Sloop's chapter on Recycling, readers will encounter a detailed discussion of battery chemistry and how it relates to recycling. Once again, an industry expert reviews the choices that designers make in battery composition and how they relate to establishing the all-important circular economy for power sources.

- Repurposing for the Second Life directly relates to the circular economy. As pointed out earlier in this chapter, EV batteries retain significant utility long after their use in EVs is complete. Drawing from the latest research in this burgeoning field, Apoorva Roy, Hamidreza Movahedi, and Anna Stefanopoulou provide an overview of the many new considerations that come into play when a vehicle's primary power source can be transitioned from propulsion to storage.

- What do engineers, managers, and other business professionals need to think about as they create and manage teams to work within this new and dynamic ecosystem? John Warner explores this in Designing New Engineering Teams and Practices. Building safe, cost-efficient, and reliable EVs is not just a technical matter. It is also a people matter. Resourcing, training, and managing those people are all critical to success. John Warner guides readers through the risks and options associated with EV and battery development.

At the very end of this book, we collect, expand, and organize the job roles mentioned in the previous chapters and serve as a guide to readers of the type of jobs in this field, the required skills, and training.

Words to Know

Black mass   A waste product consisting of shredded ingredients such as lithium, cobalt, and other rare earth materials.

Brine   The liquid materials containing lithium and other materials that can be refined for use in batteries. It is extracted from beneath the ground.

Circular economy   A process that incorporates the reuse of materials, with the goal of producing as little waste as possible.

Energy transfer within the context of mobility   The transition from fossil fuel-powered propulsion to electrification.

Lifecycle assessment (LCA)   It codifies the steps in product manufacturing and quantifies their environmental impacts.

Paris Agreement   An international treaty aimed at mitigating climate change and was signed in 2015.

Second Life within the context of battery development   The use of the battery after it fulfills its primary use.

Sustainability   Meeting a civilization's present needs without compromising the ability of future generations to meet their own needs, achieved through the responsible management of resources and social systems.

Further Reading

Here are publications we have found useful in building a general understanding of sustainability and its role in many aspects of mobility discussed in this chapter.

Frankopan, P. (2023). *The Earth Transformed: An Untold Story*. New York: Random House A comprehensive history of how Earth's climate has changed since pre-history, providing valuable context for the changes we are undergoing.

Henderson, R. (2020). *Reimagining Capitalism in a World on Fire*. New York: PublicAffairs An excellent examination of business and leadership in the context of sustainability with in-depth examinations of companies, such as Unilever, that have succeeded in balancing financial growth with social responsibility.

McKibben, B. (2020). *Falter: Has the Human Game Begun to Play Itself Out?* New York: Holt The founder of the influential 360.org examines the best and worst case scenarios in global warming.

Meadows, D. (2008). *Thinking in Systems: A Primer*. Junction, VT: Chelsea Green Meadow's overview of system thinking as a means solving problems in both social and physical infrastructure has influenced approaches to sustainability.

Pohlman, P. and Winston, A. (2021). *New Positive: How Courageous Companies Thrive by Giving More than They Take*. Cambridge, MA: Harvard Business Review Press The former president of Unilever and how to lead a sustainable company.

Sachs, J. (2015). *The Age of Sustainable Development*. New York: Columbia University Press An overview of the subject of sustainable development by the director of the Earth Institute.

Smil, V. (2022). *How the World Really Works*. New York: Viking Offers excellent examples of how quantifying the environmental impact of a product, such as tomatoes, can lead to unexpected insights. Also conveys the complexity behind terms like decarbonization.

Von Bertalanffy, L. (2015). *General System Theory Revised Edition*. New York: George Braziller A revision of the foundational work that established systems thinking, a holistic view of the interactions between systems in nature and society that informs sustainability.

Wallace-Well, D. (2020). *The Uninhabitable Earth: Life After Warming*. New York: Crown A stark look at the facts of global warming and their potentially dire consequences.

Yergin, D. (2021). *The New Map: Energy, Climate and the Clash of Nations*. New York: Penguin Books An examination of how power generation – particularly in the transportation industry – drives geopolitics.

CHAPTER 1

High-Capacity Battery Technology

David Howell
Strategic Marketing Innovations (formerly with the Department of Energy), Washington, DC, USA

Tien Duong
Department of Energy, Vehicle Technologies Office, Washington, DC, USA

Introduction

Lithium batteries have come to dominate the battery market powering devices ranging from cell phones to electric vehicles (EVs) and becoming the linchpin to ensure the reliability of the increasing renewable grid. The lithium-ion (Li-ion) battery (LIB) is a platform technology, with many different anodes, cathodes, and electrolyte combinations that change the fundamental characteristics of the electrochemistry and lead to different performance metrics.

This chapter describes the various battery component active materials that are currently used in commercial LIBs and their characteristics. In addition, emerging Li-ion concepts are examined with an eye on the advantages and challenges that need to be solved for adoption into EVs. The chapter also covers the emerging Lithium metal-based battery materials and sodium-ion battery technology, that offer a variation to Li-ion but with significantly better supply chain security.

What You Will Learn in This Chapter

The most important learning from this chapter is that the diversity of lithium battery materials has led to variants for EVs and grid storage applications, with some chemistries promising long driving ranges while others lead to lower costs. The reader will also learn:

- The basic construction of LIBs.
- About the trade-offs that exist in energy density, cost, and commercial availability.

Electric Vehicle Batteries: From Sourcing to Second Life and Recycling, First Edition.
Edited by Bob Galyen and Frank Menchaca.
© 2025 John Wiley & Sons, Inc. Published 2025 by John Wiley & Sons, Inc.

- About the supply chains associated with specific battery components.
- About the latest research trends to find sustainable, energy-dense alternatives to more traditional rare earth materials.

Lithium-Ion Battery Technology

The remarkable success of the LIB is a result of simultaneously achieving high-energy density, reasonable power density, and long cycle and calendar life. This success is due to the reliance on the concept of intercalation, with lithium ions stored in the host lattice leading to minimum transformation of the structure and achieving long cycle life. Since the commercialization of the LIB started three decades ago, the materials used for the anode, cathode, and electrolyte have evolved in response to the need for higher-performance batteries for consumer electronics, EV, and grid storage applications. Figure 1.1 illustrates the material structure of a graphite-transition metal oxide battery popular in many applications.

Present Status of Li-Ion Battery-Active Materials

Many different anode and cathode materials have been examined aimed at enhancing the various performance metrics in the battery. Higher energy density requires materials that operate at high cell voltage along with high capacity. However, the choice of such materials needs to be balanced with the elevated risk of side reactions as cell voltage increases and possible structural changes in the material with increasing lithium content (i.e. capacity). Electrolytes play an important role because most battery materials operate above the thermodynamic limit of the electrolytes. Battery R&D aims to develop or discover new materials that increase energy density without compromising on other metrics, such as cycle life and safety.

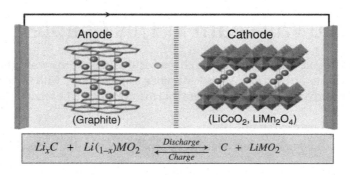

FIGURE 1.1 Schematic of a typical Li-ion battery with the electrode reactions

TABLE 1.1 Status of Various Li-ion Materials and Their Suitability Across Different Metrics

Couple	Cell Voltage (V)	Specific Energy (Wh/Kg)	Cycle Life	Fast Charge	Thermal Stability	Representative Manufacturer	Application
Graphite|| LiFePO$_4$	3.2	180–200	Excellent	Good	Excellent	CATL, BYD	Tesla Model Y
Graphite|| LMFP (LiMn$_{1-x}$Fe$_x$PO$_4$)	3.8–4.0	220–230	Good	Good	Excellent	Saft, CATL, BYD	EVs and others
Graphite|| LiMn$_2$O$_4$ + NMC	3.7–3.75	220	Fair	Fair	Good	LG-Chem	Chevy Volt
Graphite|| NCA	3.65	200–260	Good	Fair	Good (gassing)	Tesla	Tesla Model S, X, 3 and Y
Graphite|| NMC	2.4	90–100	Excellent	Excellent	Good	Microvast	Buses
Graphite|| NMC (Various Composition of Ni)	3.6	220–260	Good	Fair	Good (gassing)	LG-Chem, Samsung, SK	Chevy Bolt, Match-E
Graphite + Silicon Composite || NMC (High Nickel)	3.5	260–320	Fair	Good	Good (gassing)	Gotion, Zenlabs	Drones

Cathode materials: While a similar interplay between voltage and cyclability also plays into the choice of cathode material, an additional consideration is the need to minimize the use of critical elements that suffer from insecure supply chains. While the layered lithium cobalt oxide dominates the consumer electronics market, the high price of cobalt has led to the commercialization of nickel–manganese–cobalt oxides and nickel–cobalt–aluminum oxide for electric vehicle applications. Recently, there has been a push toward using the olivine, lithium iron phosphate, and variants that incorporate manganese, as an alternative to cathodes with nickel and cobalt. The cathode has an advantage in high stability and safety, owing to its lower voltage, but suffers from lower energy density compared to the oxide analogies. Table 1.1 tabulates the status of the different cathode materials used in commercial LIBs across the different metrics along with representative manufacturers that have commercialized the technology.

With the increasing concern over critical material use in batteries, R&D efforts have focused on reducing, and ultimately eliminating, cobalt and possibly nickel in next-generation batteries. The present focus is on reducing cobalt content by adding more nickel to the cathode along with other substitutions such as iron and aluminum. This trend is part of the movement from 33% cobalt (NMC-333 cathodes) to 20% (NMC-622) to the latest generation of cells that aim to use 10% or less cobalt (NMC-811 and NMC-90505) by substituting more nickel to the lattice. While higher nickel content promises higher capacity and thereby higher energy density, it comes at the expense of increased reactivity and lower thermal stability due to more propensity for oxygen release during thermal events. Differential scanning calorimetry data (Figure 1.2) shows the trend toward lower temperatures for the release of heat as the nickel content increases accompanied by higher heat release. NMC-811 cathodes are also prone to particle cracking at higher voltages due to the molar volume change in the lattice, resulting in a movement toward single-crystal cathodes.

Anode materials: The most prevalent type of LIB anode is a carbon graphite anode and can be produced from natural graphite, synthetically produced graphite, or a mix of both. Optimized graphite anode materials typically have a capacity of >350 mAh/g (close to the theoretical value of 372 mAh/g for LiC_6). While current LIBs are cathode limited, improvements in anodes continue to drive up energy

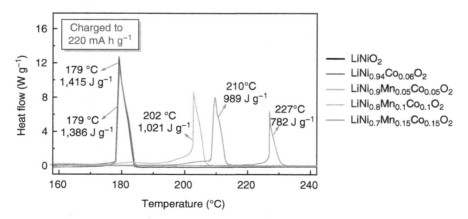

FIGURE 1.2 Differential scanning calorimetry (DSC) data on various TM cathodes with changing Ni content. *Source: Figure provided by Prof. Arumugam Manthiram, U. of Texas at Austin.*

density and aid in decreasing cost. Silicon composite anodes have improved significantly in the last decade. Commercial cells now blend low amounts of silicon with graphite (typically <10%), resulting in enhanced capacity and performance.

PRACTICAL INSIGHTS | Using Silicon for Battery Anodes

The same material used in pencils (graphite) has long been a key component in today's lithium-ion batteries LIBs. As reliance on these batteries increases, however, graphite-based electrodes are due for an upgrade.

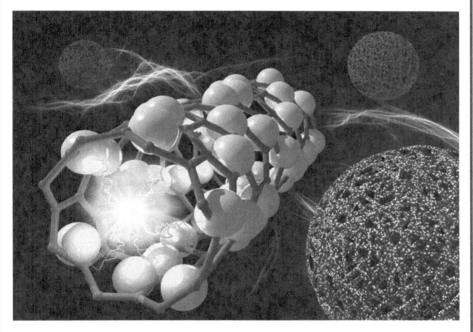

Silicon microspheres have extraordinary mechanical strength due to the addition of carbon nanotubes, which make the spheres resemble balls of yarn. In this representation, the image on the left illustrates a close-up of a portion of a microsphere made of silicon nanoparticles deposited on carbon nanotubes. *Source: Mike Perkins, Pacific Northwest National Laboratory.*

Silicon, used in computer chips and many other products, is appealing because it can hold 10 times the electrical charge per gram compared to graphite; however, silicon expands greatly when it encounters lithium and is too weak to withstand the pressure of electrode manufacturing. Researchers have developed a unique nanostructure that limits silicon's expansion while fortifying it with carbon. This work could inform new electrode material designs for other types of batteries and eventually help increase the energy capacity of LIBs in electric cars, electronic devices, and other equipment.

A conductive and stable form of carbon, graphite, is well suited to packing lithium ions into a battery's anode as it charges. Silicon can take on more lithium than graphite but it tends to balloon about 300% in volume, causing the anode to break apart. The

(continued)

> **PRACTICAL INSIGHTS** | Using Silicon for Battery Anodes
> (*continued*)
>
> researchers created a porous form of silicon by aggregating small silicon particles into microspheres about 8 micrometers in diameter – roughly the size of one red blood cell.
>
> The electrode with a porous silicon structure exhibits a change in thickness of less than 20% while accommodating twice the charge of a typical graphite anode. Unlike previous versions of porous silicon, the microspheres also exhibited extraordinary mechanical strength, thanks to carbon nanotubes that make the spheres resemble balls of yarn.
>
> The researchers created the structure in several steps, starting with coating the carbon nanotubes with silicon oxide. Next, the nanotubes were put into an emulsion of oil and water. Then they were heated to boiling. The coated carbon nanotubes condense into spheres when the water evaporates. Then, aluminum and higher heat were used to convert the silicon oxide into silicon, followed by immersion in water and acid to remove by-products. What emerges from the process is a powder composed of tiny silicon particles on the surface of carbon nanotubes.
>
> The porous silicon spheres' strength was tested using the probe of an atomic force microscope. One of the nanosized yarn balls may yield slightly and lose some porosity under very high compressing force but it will not break. Anode materials must be able to handle high compression in rollers during manufacturing.
>
> The next step is to develop more scalable and economical methods for making the silicon microspheres so that they can one day make their way into the next generation of high-performance LIBs.
>
> Source: "Using Silicon for Battery Anodes," Mobility Engineering, October 1, 2020.

Next Generation Li-Ion Cathode and Anode Research and Development

Next-generation Li-ion chemistries employ an alloy anode that is normally silicon-based and/or a high-voltage and high-energy cathode. These cells promise 20–40% higher energy density than today's cells, potentially lower cost, and reduced dependence on critical battery materials.

Manganese-rich NMC cathodes: Lithium- and manganese-rich (LMR) oxides promise several advantages over current state-of-the-art (e.g. Ni-rich) cathodes as they are typically made up of ~50% or more manganese. Specifically, manganese is known to enhance the stability of cathodes in terms of safety and is one of the world's most abundant, readily available (geographically diverse), and inexpensive transition metals. Data on these materials (see Figure 1.3) (Croy et al. 2021) suggests compelling performance compared to industrially prepared NMC-622 cycled under standardized, high-voltage protocols. Although promising, several challenges remain that inhibit large-scale adoption. These include (1) mitigating impedance at low states of charge, (2) enhancing surface instability over long-term cycling, (3) inhibiting or controlling local cation rearrangements, and (4) enabling particle and electrode designs that allow for higher volumetric energies.

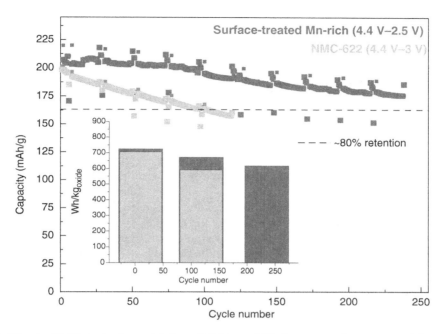

FIGURE 1.3 LMR electrode containing ~60% Mn and 0% Co compared to commercial NMC-622. *Source: Croy et al. (2021)/U.S. Department of Energy's Office of Science.*

***Disordered rocksalt* cathodes:** While reducing cobalt has been a trend that is decades in the making, the recent price increases for nickel have led to concern about the long-term viability of substituting nickel for cobalt. This has led to a push to reduce nickel content by moving toward cathodes that consist exclusively of earth-abundant materials. A promising recent example is the research on Li-excess disordered rocksalt cathodes (or DRX), a new class of materials that are largely unexplored and have broad chemical flexibility. DRX materials with high Mn content and fluorine substitution display promising thermal and cycling stability, along with high specific energy (Ceder 2023). Figure 1.4 shows one such example of the excellent cycleability of the DRX cathode. Rate capability has been enhanced from earlier DRX materials through careful optimization of composition and synthesis. Active research on DRX materials focuses on optimizing the sloppy voltage profile and matching the DRX chemistry with high-voltage stable electrolytes. Due to their potential as earth-abundant inexpensive cathode materials, they are an active area of research.

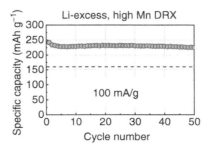

FIGURE 1.4 Cycling stability of DRX cathode containing only earth-abundant Mn and Ti. *Source: Ceder (2023)/U.S. Department of Energy's Office of Science.*

In terms of performance and stability, DRX cathodes based on Mn-rich oxyfluorides with F content ranging between 0.1 and 0.2 with nominal composition, $Li_{1.2}Mn_{0.6}Ti_{0.2}O_{1.8}F_{0.2}$ (LMTF2622) and $Li_{1.2}Mn_{0.7}Ti_{0.1}O_{1.9}F_{0.1}$ (LMTF2711) show great promise. Optimizing the effective voltage window and using carbon and other inorganic surface coatings to enhance electronic conductivity and stability enabled stable capacity retention of >220 mAh/g exceeding 300 cycles. Some compositions, such as the high-Mn DRX under extensive electrochemical cycling, undergo a structural transformation to the δ phase. This transformation leads to capacity gain in early cycles and increases with Mn content, as shown in Figure 1.5 (Comparison between 1st cycle and 30th cycle) (Ceder 2024/U.S. Department of Energy Office of Science). Increasing Mn content also leads to faster and more extensive δ phase transformation.

Recent progress in synthesis and processing methods has led to ultra-high Mn DRX materials with nominal composition, $Li_{1.1}Mn_{0.8}Ti_{0.1}O_{1.9}F_{0.1}$. Design of targeted high-voltage electrolytes for DRX cathodes would further enable capacity retention and ensure interfacial stability at the top of the charge.

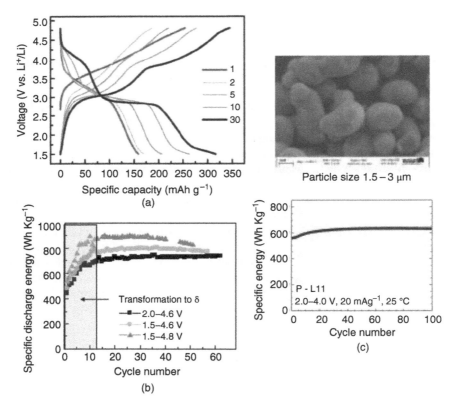

FIGURE 1.5 (a) Transformation from pristine DRX to δ phase with cycling for $Li_{1.1}Mn_{0.8}Ti_{0.1}O_{1.9}F_{0.1}$, (b) Specific discharge energy of Gen-3 DRX with varying voltage window number of cycles to transform to δ phase, and (c) Single Crystal Gen-3 DRX with stable cycling and controlled particle size distribution. *Source: Ceder (2024) / Lawrence Berkeley National Laboratory / public domain.*

Case Study: Researching Cobalt Alternatives

The batteries that power our cell phones, laptops, and EVs rely on cobalt. Cobalt mining is problematic for the environment and miners. In order to find other solutions for lithium-ion (Li-ion) batteries that move away from a dependency on cobalt, researchers at the U.S. Department of Energy's (DOE) Argonne National Laboratory participated in a collaborative study to identify new potential materials for the positive terminal of a battery, called a cathode.

In a battery, Li-ion is inserted into a cathode during charging and released during discharging, providing electricity. The new cathodes offer two advantages: They are both cobalt-free and stable, which means that they do not undergo a structural failure, such as cracking, as they are repeatedly charged and discharged.

As a battery constituent cobalt offers thermal stability, which means it functions even as it is heated to higher temperatures, as well as structural stability. Researchers have been looking for different materials that could offer these same advantages without cobalt's flaws.

In a 2022 study, a research team led by the University of California, Irvine, created and analyzed a material for a Li-ion cathode that uses no cobalt and is instead rich in nickel. This cathode chemistry is compositionally complex, meaning that it contains small amounts of a wide range of other metals. These metals include molybdenum, niobium, and titanium.

"You can think of building a cathode like building a house out of different kinds of bricks," said Argonne Physicist Wenqian Xu. "By having a variety of different shapes and sizes of bricks, we can enhance the stability of the house. Multiple elements help to ensure the integrity of the cathode particles."

The researchers wanted to investigate the structural and thermal stability of the new cathode. Other nickel-rich cathodes typically have poor heat tolerance, which can lead to oxidization of battery materials and thermal runaway, which could in some cases lead to explosions. Additionally, even though high-nickel cathodes can accommodate larger capacities, large changes in volume from repeated expansion and contraction can result in poor stability and safety concerns.

To test the new battery, the researchers cycled it more than a thousand times. They discovered that in the process, the cathode material underwent less than 0.5% of volume expansion. This is roughly a tenth of the volume expansion experienced by previous nickel-rich cathodes, which all had stability problems to varying degrees.

"Keeping the volume of the cathode consistent is essential for ensuring its stability," said Argonne Physicist Tianyi Li.

To characterize the heat tolerance of the new cathode material, called HE-LMNO, the UC Irvine team used beamline 11-ID-C at the APS, with the support of Xu and Li, to examine what would happen to the material at high temperatures. As opposed to previous high-nickel cathodes, which showed severe nanocracking at high temperatures, the HE-LMNO undergoes a phase change that allows it to continue to perform and retain capacity. The HE in HE-LMNO stands for high entropy, a characteristic that refers to the large number of different elements included in the alloy.

"The APS significantly advanced our understanding of the high-entropy doped material we studied," said UC Irvine's Huolin Xin, the lead author of the study.

(continued)

> **Case Study: Researching Cobalt Alternatives (*Continued*)**
>
> "Our results suggest the high-entropy effect is transferable to a broader class of compounds that could form the basis of new battery materials."
>
> According to Xu, the research could provide design rules for a host of new battery cathodes that could help reduce next-generation LIBs' reliance on cobalt. "We haven't just found one new battery," he said. "Really, by mixing different transition metals in the structure, we could potentially see many more interesting cathode candidates."
>
> Source: "New Lithium-Ion Battery Cathode Offers Higher Stability," Mobility Engineering, September 1, 2023.

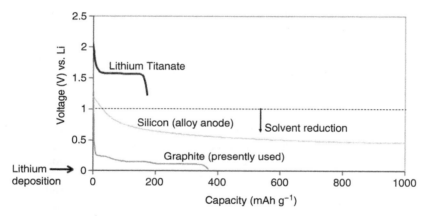

FIGURE 1.6 Voltage vs. capacity curves for various anode materials.

Anode materials: In the case of anodes, silicon-based anodes have been researched as an alternative to graphite due to their significantly higher capacity. However, this increase comes at the expense of significant volume change (as much as 300% change per cycle), which leads to stress buildup and particle cracking. While the push to increase energy density favors the use of anodes with low voltage, these come at the peril of side reactions, such as SEI formation and lithium plating during the fast charge of graphite anodes. Attempts have been made to move toward higher voltage anodes (which leads to lower cell voltage and therefore lower energy density) to improve the cycle life. Silicon anodes have improved significantly in the last decade with commercial cells blending silicon in graphite with increasing amounts. While the cycle life of silicon has now improved to meet EV targets, the calendar life remains a significant concern, driven by the lack of a stable anode/electrolyte interface. Research continues to improve the calendar life of anodes with high silicon content. An example is the lithium titanate anode that operates at 1.5 V vs. Li which promises long life for grid applications. Figure 1.6 shows voltage curves for a few different anode materials along with the SEI formation and Li deposition voltage.

Lithium-Metal Battery Technology, Including Lithium–Sulfur and Solid-State Materials

Achieving significant cell level capacity increases is not likely with current battery chemistries. Lithium metal as the anode has long been the dream for battery researchers, due to its high capacity and low anode voltage. The high-energy density of a thinly protected lithium metal anode needs to be coupled, either in the all-solid-state or with a purpose-designed liquid catholyte, with a cathode active material capable of substantially more than lithium-intercalating oxides can bring significant capacity gains. However, lithium's high reactivity leads to the formation of dendrites during charge and enhanced side reactions with the electrolyte. These lead to capacity fade and loss of cyclable lithium, in addition to the propensity for cell shorting and associated thermal events. Controlling dendrites remains a challenge for this promising anode material.

Li-metal anodes: Li-metal anodes have also become an active area of research with the emergence in the last decade of promising super-ionic single ion conducting inorganic materials that show promise in mechanically preventing dendrites and limiting reactivity. These span materials are based on oxides, sulfides, and halides. Each type of material exhibits favorable attributes over some metrics, while suffering in others, as illustrated in Table 1.2 (Duong 2021). Efforts continue to solve

TABLE 1.2 Status of Solid Electrolytes Under Study Spanning Organic and Inorganic Materials

Representative:	Polymer PEO	Oxides LIZO	Sulfides Li_2S-nP_2S_5 Blends	Halides Li_3YCl_6, Li_3InCl_6
Material phase	Amorphous	Crystalline	Crystalline or glass	Crystalline
Ionic conductivity	Poor	Fair	Good	Good
Air stability	Good	Good	Poor	Poor
Stability against Li anode	Good	Good	Poor	Poor
Stability against high-V cathode	Fair	Good	Poor	Good
Ease of manufacturing/ processing technique	Good/ roll-to-roll	Fair/ sintering	Good/ roll-to-roll	Too early
Stack pressure required	Image	Image	Image	Image
Companies	Hydro Quebec, Bellore, Seeo	Iron storage system, Quantumscape	Toyota, Samsung, Solid Power, PolyPlus	None at this time

Source: Duong (2021)/U.S. Department of Energy's Office of Science.

the challenges in inorganic materials. Polymer-based materials based on polyethylene oxide (PEO) have been under study for decades, with commercial cells showing some success due to the issues including the need for high temperatures. Efforts are also underway to combine polymer and inorganic materials to enable composites to attempt to combine the best aspects of these materials. In addition to solids, improvements in liquid electrolytes (Liu 2021/Pacific Northwest National Laboratory), such as the emergence of localized high-concentration electrolytes, have shown promise in enabling stable cycling at high-energy densities with Li-metal. Figure 1.7 shows an example of the excellent cycling of Li-metal anode with such advances. A significant challenge remains including preventing dendrites on the fast charge, calendar life issues, and the safety of cells with Li metal with solid electrolytes playing an increasingly important role. Further solid-state batteries also introduce manufacturing constraints depending on the choice of materials, with hard inorganics being especially vulnerable to defects during processing. Defects become nucleation points for dendrite growth. Another important area is the need for thin, low-cost lithium sources to ensure that there is limited excess lithium in the cell to maximize energy density. Large-scale production of such a thin anode remains a challenge.

Li metal remains an area of active R&D since the anode will serve as a platform to incorporate high-energy cathodes such as sulfur and oxygen – important innovations that will further improve energy density and enable electrification of the transportation sector.

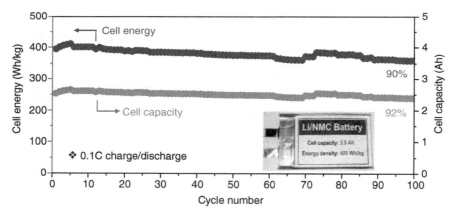

FIGURE 1.7 Cycling of Li metal/liquid/NMC cathode pouch cells. *Source: Liu (2021)PNNL/U.S. Department of Energy/Public Domain.*

PRACTICAL INSIGHTS | A Deeper Look at Lithium Metal Anodes

In the pursuit of a rechargeable battery that can power EVs for hundreds of miles on a single charge, scientists have endeavored to replace the graphite anodes currently used in EV batteries with lithium metal anodes. However, while lithium metal extends an

EV's driving range by 30–50%, it also shortens the battery's useful life due to lithium dendrites – tiny, treelike defects that form on the lithium anode over the course of many charge and discharge cycles. Dendrites also short-circuit the cells in the battery if they make contact with the cathode.

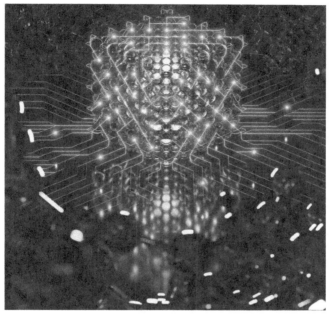

Researchers have designed new solid electrolytes that light the path to wider electrification of transportation. *Source: Courtesy of Jinsoo Kim.*

For decades, researchers assumed that hard, solid electrolytes, such as those made from ceramics, would work best to prevent dendrites from working their way through the cell. But the problem with that approach is that it didn't stop dendrites from forming or nucleating in the first place, like tiny cracks in a car windshield that eventually spread.

Researchers have developed a new class of soft, solid electrolytes – made from both polymers and ceramics – that suppress dendrites in that early nucleation stage before they can propagate and cause the battery to fail.

Solid-state energy storage technologies such as solid-state lithium metal batteries that use a solid electrode and a solid electrolyte can provide high-energy density combined with excellent safety but the technology must overcome diverse materials and processing challenges. The new dendrite-suppressing technology could enable battery manufacturers to produce safer lithium metal batteries with both high-energy density and long cycle life. Lithium metal batteries manufactured with the new electrolyte could also be used to power electric aircraft.

Key to the design of the soft, solid electrolytes was the use of soft polymers of intrinsic microporosity (PIMs) whose pores were filled with nanosized ceramic particles. Because the electrolyte remains a flexible, soft, solid material, battery manufacturers will be able to manufacture rolls of lithium foils with the electrolyte as a laminate between the anode and the battery separator. The lithium-electrode sub-assemblies

(continued)

> **PRACTICAL INSIGHTS** | A Deeper Look at Lithium Metal Anodes (*continued*)
>
> (LESAs) are attractive drop-in replacements for the conventional graphite anode, allowing battery manufacturers to use their existing assembly lines.
>
> To demonstrate the dendrite-suppressing features of the new PIM composite electrolyte, the researchers created 3D images of the interface between lithium metal and the electrolyte to visualize lithium plating and stripping for up to 16 hours at high current. Continuously smooth growth of lithium was observed when the new PIM composite electrolyte was present, while in its absence, the interface showed telltale signs of the early stages of dendritic growth. These and other data confirmed predictions from a new physical model for electrodeposition of lithium metal.
>
> Source: "Battery Improvement Enhances Electric Flight and Long-Range Electric Cars," Mobility Engineering, October 1, 2020.

Sulfur cathodes: Enabling lithium metal brings the possibility of transitioning cathodes away from transition metals toward high energy alternates such as sulfur and oxygen which are earth-abundant. Li-S batteries promise to increase the gravimetric energy significantly, with theoretical values (2600 Wh/kg) exceeding current Li-ion (400 Wh/kg) by a factor of six. Technical feasibility of lithium–sulfur batteries has been made difficult by several materials and cell design issues, including the polysulfide shuttle phenomenon, volume change with cycling, and low electronic and ionic conductivity, which lead to poor cycling behavior and low-energy density due to low S utilization in high loading electrodes. To date, only modest energy densities and low cycle life cells have been achieved due to the difficulty of increasing sulfur utilization in high-loading sulfur cathodes. Because the sulfur cathode depends on the conversion of sulfur to Li_2S on discharge and the reverse reaction during charging, lithium ions and electrons must be distributed to the sulfur cathode active material to enable the reaction, as sulfur is neither electronically nor ionically conductive. Furthermore, the complex set of electrochemical steps and intermediates between elemental sulfur, S_8, and Li_2S, impose thermodynamic, kinetic, and transport limitations (including the high solubility of the intermediate polysulfide species in organic electrolytes).

In the near term, the use of sulfurized polyacrylonitrile (SPAN), formed by hearing polyacrylonitrile with sulfur under inert conditions, promises to stop the polysulfide shuttle, thereby enhancing cycle life. Compared to the conventional carbon–sulfur composite, SPAN provides better dispersion and confinement of the sulfur-based active species via means of covalent interaction. But this comes with the use of extra weight and volume, leading to a projected energy density of 250 Wh/kg. While the use of elemental sulfur can lead to energy densities closer to 500 Wh/kg, this comes at the expense of cycle life. Both approaches show promise toward enabling high-energy batteries. Combining carbon-sulfur composites with SPAN (CS-SPAN) can lead to enhanced capacity while minimizing shuttle mechanisms (Liu 2024), as shown in Figure 1.8. This approach, though preliminary, shows promise.

Oxygen cathodes: Lithium–oxygen batteries have one of the highest theoretical energy densities of any battery (3,500 Wh/kg), approaching gasoline engines. However, the chemistry remains challenging on multiple fronts, including side reactions

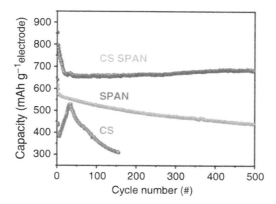

FIGURE 1.8 Capacity vs. cycle life of SPAN and CS SPAN-based Li-S cells. *Source: Liu (2024)/ Pacific Northwest National Laboratory.*

in the electrolytes, reactions with impurities in an open system exposed to ambient air, formation of insulating layers, etc. While closed systems circumvent some of these challenges, technoeconomic models suggest that the energy density decreases drastically under this scenario. Recent advances in the use of catalytically active supports, along with ionic liquid electrolytes show promise enabling a solution-mediated reaction to achieve high discharge capacity. While these studies are still preliminary, they suggest that Li-air batteries warrant a fresh look, with a focus on new cathode materials and nanostructures, anode protection schemes, and novel electrolytes (both liquid and solid). Enabling such chemistry can be significant in electrifying the transportation sector.

Sodium-Ion Battery Technology

With the increasing concern about the availability of lithium, there is growing interest in the diversity of batteries with other working ions including protons, sodium, magnesium, calcium, and zinc. Of these variations, sodium-ion (Na-ion) batteries have garnered a lot of interest recently for both transportation and grid applications. Despite the lower cell voltage when using sodium leading to lower energy density, promising new materials have been found that may allow this chemistry to become a viable alternative to lower-energy LIBs, such as LFP-based cells.

In the area of anodes, graphite, the dominant anode for LIBs, is incompatible with Na-ions, and thus alternative anodes must be explored. Instead, hard carbon anodes have become more prominent. While chemically like graphite, it lacks the ordered structure and interacts with Na+ through adsorption, intercalation, and pore-filling processes. This leads to a slopping voltage profile starting near 1.2 V and down to nearly 0 V vs. Na/Na+. Typical capacities range from 150 to 350 mAh/g with capacities as high as 478 mAh/g reported. Poor first-cycle efficiency, inconsistency in materials, and low potentials leading to Na plating and dendrite formation are key areas that must be addressed for improved performance and reliability. In addition, alloys, such as tin and phosphorus, are also being examined due to the potential for high capacity. One major concern is the volume expansion of the alloy that negatively impacts the cycle life. Similar to Li metal, Na metal offers low voltages (−2.71 V vs.

Standard Hydrogen Electrode [SHE]) and high capacities (1166 mah/g) which would enable high-energy dense batteries. Na metal is even more reactive than Li metal and this only amplifies the safety concerns that already exist with Li metal anodes. Due to this, the use of Na metal for EV applications remains problematic.

In the area of cathodes, NIBs benefit from an increased variety of stable cathode structures compared to LIBs due to the larger ionic radius of Na-ions. However, simply transitioning from LIB cathodes to NIB systems (e.g. $LiCoO_2$ to $NaCoO_2$) results in drastically different voltage profiles (increased sloping and step behavior). Stronger Na–Na interactions further increase the complexity of NIB cathodes. Three classes of materials that have gained prominence include Prussian blue (PB), layered oxides, and polyanions.

PB cathodes are a class of iron hexacyanoferrate ($Na_xM_y[Fe(CN)6]_z$). Their lower voltage (3.5 V) and capacities (140–160 mAh/g) limit the cell energy density to only 80 mAh/g. Further issues with the difficulty of removing water from the structure, toxicity, difficulty in producing, and challenges in making reproducible materials have all limited the success of PB cathodes. Despite these downsides, PB has found success commercially due to the large sites for strain-free insertion of Na-ions, cheap cost, and flatter voltage profile when compared to other cathode materials.

Layered oxide promises higher capacity compared to other cathode types for NIB, making them a promising avenue for exploration, such as LIBs. While LIBs have relied heavily on Ni, Mn, and Co for the layered oxides (Ni and Co being critical materials with cost and supply chain concerns), NIBs have employed a wider variety of compositions including Al, Mn, Ni, Fe, Cu, Mg, Co, Ti, Li, etc. This is seen as advantageous both in terms of the flexibility to deal with supply chain issues and to provide ample space to optimize chemistries for specific applications. Current challenges are in stabilizing the voltage profile while maintaining high-energy density.

Polyanions represent a growing area for NIBs, especially for grid applications. These cathodes share many of the same transition metal components as found in layered oxides, but where they differ is in the use of multielement anions (e.g. SO_4^{2-}, PO_4^{3-}, $[P_2O_7]^{4-}$) as opposed to the mono-element anion of the oxides (O_2^{4-}). Polyanion cathodes push the redox potential higher (up to 4.7 V) but do so at the cost of capacity and electronic conductivity. One of the most popular polyanion chemistries is sodium vanadium phosphate (NVP) and the fluoride (NVPF) variant. While these cathodes have shown some success in research and are being utilized commercially, the use of vanadium is a major barrier due to its high cost and sourcing concerns (the top two producers in 2019 were China and Russia). This cathode class is a promising alternative to layered oxides, but considerable research needs to be invested into improving the energy density of these materials as well as ensuring that they do not rely on critical materials.

Future Considerations

The growing interest in electrifying the transportation sector has led to increased R&D in battery technology. The LIB has improved constantly over the past three decades to enable widespread adoption in many applications.

The future looks bright with the emergence of new materials for anodes, cathodes, and electrolytes for next-generation Li-ion and Li-metal-based batteries. However, numerous challenges remain that need long-term sustained R&D to ensure the development of next-generation, high-energy, cost-effective, and safe batteries that can usher in an era of electrified transportation.

Words to Know

Intercalation   The process by which a molecule is inserted into a lattice structure; this process defines the chemical process that occurs when a battery is charging or discharging.

Differential scaling calorimetry   A method for measuring a battery material's heat capacity during heating or cooling as a means of evaluating its performance.

Further Reading

The Vehicle Technologies Office at the U.S. Department of Energy provides links to research and development of battery technologies. **https://www.energy.gov/eere/vehicles/batteries**
The SLAC-Stanford Battery Center at Stanford University also focuses on advanced battery technology research and industrial application. **https://batterycenter.slac.stanford.edu/research**

References

Ceder G. (2023). *Cation-disordered cathode materials(DRX+): Overview and Progress,* **https://www1.eere.energy.gov/vehiclesandfuels/downloads/2023_AMR/bat570_ceder_2023_o%20-%20Gerbrand%20Ceder.pdf** (accessed 17 January 2025).

Ceder G. (2024). *Cation-disordered cathode materials(DRX+): Overview,* **https://www1.eere.energy.gov/vehiclesandfuels/downloads/2024_AMR/BAT570_Ceder_2024_p.pdf** (accessed 17 January 2025).

Croy J. et al. (2021). *High-Capacity Cathodes Based in Earth-Abundant Manganese: Scratching the Surface. . . Or Not?* **https://www.anl.gov/access/article/highcapacity-cathodes-based-in-earthabundant-manganese-scratching-the-surfaceor-not** (accessed 17 January 2025).

Duong T. (2021), *Overview of Battery Materials Research Program,* **https://www.energy.gov/sites/default/files/2021-06/bat108_duong_2021_o6_6.8_10.08am_LS.pdf** (accessed 17 January 2025).

Liu J. (2021), *Overview and status of Battery500 consortium,* **https://www.energy.gov/sites/default/files/2021-06/bat317_liu_2021_p_5-12_735pm_LR_TM.pdf** (accessed 17 January 2025).

Liu J. (2024), *Progress and status of Battery500 consortium Phase II,* **https://www1.eere.energy.gov/vehiclesandfuels/downloads/2024_AMR/BAT317_Liu_2024_o.pdf** (accessed 17 January 2025).

CHAPTER 2

Sourcing Lithium-Ion Batteries

Austin Devaney
Li7Charged, Brevard, NC, USA

Bob Galyen
Galyen Energy LLC, Noblesville, IN, US

Introduction

The lithium-ion battery (LIB) has become the cornerstone of the electrification movement. Its future outlook and projected growth show that it is at the center of a new Industrial Revolution. LIBs necessitate the creation of a new supply chain network to support a Terawatt level of energy storage.

Such significant growth brings the concern about supply shortages of materials necessary to meet the demand for electric vehicle (EV) adoption and energy storage, which demand significant volumes of these batteries. New strategic plans for sourcing, while balancing supply–demand dynamics in the global market, are of paramount importance. Sourcing the large volume of materials required to make batteries requires a regionalization plan and government policies supporting an international collaboration on critical materials. At the same time, reducing dependencies on single source regions of concern, which dominate the global market today, is also a top concern (Ren et al. 2024).

This regionalization plan for sourcing carries with it the burden of evolving technology for processing, transporting, and utilization within battery production plants. Localization also means getting the raw materials close to the battery producers to minimize transportation costs, improve just-in-time deliveries, and avoid natural or man-made supply blockages to those factories. At the same time, the environmental impact of pollutants on water and air streams in the processing and transport of these precious materials must be addressed (Negrete et al. 2024). Advanced extraction technologies and innovations in more sustainable mining and material processing certainly play a role in this strategy. Improvements in technologies to achieve better yields and reduce environmental impact are also important.

Electric Vehicle Batteries: From Sourcing to Second Life and Recycling, First Edition.
Edited by Bob Galyen and Frank Menchaca.
© 2025 John Wiley & Sons, Inc. Published 2025 by John Wiley & Sons, Inc.

What You Will Learn in This Chapter

In this chapter, we will examine the critical topic of sourcing battery materials, particularly within the context of sustainability. The reader will:

- Build an understanding of the supply chain for LIBs.
- Review battery cell types, particularly in the context of sourcing.
- Understand both the regional and global considerations that factor into sourcing.

Cell Types and Sourcing

In order to understand the importance of the supply chain for LIBs, we need to explain a few details of the essentials that go into making such a critical component of our society. A LIB is a modern-day dispatchable energy storage device (Figure 2.1) with great flexibility in its application.

Most LIB cells have more than 75 key components that go into the manufacturing of these sophisticated cells at the cell manufacturing plant. The LIB cell is the building block of the LIB and is mostly defined by both the packaging (which encases the active materials within the cell) and the chemistry of the cathode used (since it is the most costly component of the cell).

The types of cells include prismatic, pouch, and cylindrical. Each of these cell designs has strengths and weaknesses, which could take significant explanation, but for the sake of brevity, the following simple explanation will suffice:

- Prismatic is normally a rectangular-shaped aluminum can with a laser-welded lid that contains both termination points of the cathode and the anode.

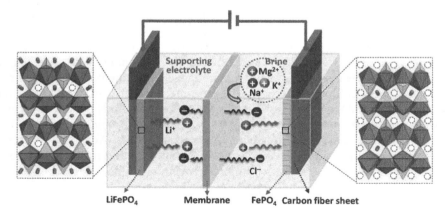

FIGURE 2.1 The structure of an electrolyte cell within a lithium-ion battery is shown. The electrolyte is the liquid within the battery structure that carries positively charged particles between the anode and the cathode. *Source: He (2018)/John Wiley & Sons/CC BY 4.0.*

- Pouch packaging is normally a soft aluminum coated with a polymer that wraps around the active materials and terminates with tabs protruding through one or two sides, which are heat sealed to contain the active materials.
- Cylindrical cell is a round cell with the termination on both ends of the cell for the cathode and the anode.
- Also, note that the key components within the cells are the cathode, anode, separator, electrolyte, and the package discussed above. Each of these key components has quite unique supply chain sources, and a variety of manufacturer technologies have pros and cons to sourcing by the cell manufacturer.

Now a brief description of the chemistries is defined by the cathode composition. Similar to packaging, chemistry has key metrics of energy density, lifespan, and cost.

- **NMC/NCA**: The most prevalent high energy density cathode has a higher price point and is known as NMC (Aluminum, Nickel, Manganese, and Cobalt Oxide).
- **LFP**: The next most common electrode is a cost-effective Lithium Iron (Ferrous) Phosphate (LFP) which has a very good cycle life but is lower in energy density than NMC.
- There are other chemistries but these two are the most common. The key point here is that each of these chemistries is unique and manufactured quite differently, creating a significant variety of materials from various technologies for processing within the supply chain network with potentially large price differences.

The importance of a robust supply chain within the battery industry cannot be overemphasized as it has strategic relevance to societal needs in the multitude of applications created within the fast-developing electrification movement. Key industries rely on the batteries made from cells of the various packaging and chemistries fueling industry demands for EV's (Figure 2.2), electronics, energy storage, and many

FIGURE 2.2 A model showing the structure of an electric vehicle. Note that the LIB is large and situated on the floor of the vehicle. Multiple cells or modules comprise the battery pack in these vehicles. *Source: visoot/Adobe Stock Photos.*

more application needs. Equally important are the defense departments around the world that see batteries as critical needs for all types of uses within military systems.

There are two main ways the cells are integrated into applications today. The first is the assimilation or grouping of cells into modules, which are then joined in a series or parallel configuration to make a battery pack for the mobility sector. Modules can be grouped into a tray or cabinet for energy storage for multiple grid support applications. The second, more common to the mobility sector, is termed cell-to-pack configuration. In this architecture, the cells are integrated directly into the pack, which commonly is a part of the architecture of the vehicle.

Global Landscape of Lithium-Ion Battery Raw Materials

Now that we have a rudimentary understanding of the form factors and chemistry connotations, we can dive a little deeper into the key materials used in making LIBs. However, before we do that, we must understand the sources, mining, processing, or production of those materials. Those most knowledgeable about the materials used in making batteries recognize that these have significant geopolitical risks, as well as environmental and ethical concerns. Some countries are restricting the export of these materials, in some cases for national security reasons and in others to block potential competitors from entering the marketplace for their own economic gain. The emergence of all these concerns has surfaced within the past 15 years with the rapid rise in volumes of larger cell formats used in electrified vehicles and energy storage systems for grid support. Therefore, driving new economies for what has become known as critical minerals in making these batteries.

Some of the major sources of these critical minerals are from countries such as Australia, South American countries, the United States, China, the Democratic Republic of the Congo, Canada, Indonesia, and many other countries. These minerals have a significant impact on the economies of those countries. However, the minerals alone are not enough to bring true value, as these minerals must be processed into precursor materials used in making the active materials for batteries. That is where real value exists. In some cases, these minerals are used within that country to make batteries or exported at a premium rate due to scarcity as the market for batteries is growing at a phenomenal rate. At present, China is the world's largest and dominant processor of raw materials to create precursor materials for making batteries. This industry was established over a two-decade period due to the large growth of consumer electronics production in that country, driven by the exponential growth in global demand for consumer electronic products. Some might view this as a prime example of resource nationalism (Li et al. 2024).

One key element in today's dominant battery chemistry is Lithium. Other elements used in making cathode materials include Aluminum, Cobalt, Nickel, Manganese, Iron, and Phosphate. In the making of anode materials, graphite, both synthetic and natural, are prevalent but the use of Silicon and metallic Lithium is entering into the designs of higher energy-rated cells. The separator industry is dominated by porous polyethylene separators, most of which are coated with Alumina for dendritic short protection. The electrolyte industry is dominated by Lithium salts dissolved into carbonate-based solvents such as Methyl Carbonate,

FIGURE 2.3 The open pit of a lithium mine. While EVs produce no tailpipe emissions, LIB creation carries a large environmental impact. Lithium mining requires large amounts of water and alters the local ecosystem and those living within it. *Source: mariiaplo/Adobe Stock Photos.*

Ethyl Carbonate, and other solvents combined with specialty additives to enhance chemical stability, improve cold and hot temperature performance, and extend the lifespan of the batteries.

For those countries developing supply chains for battery technologies, it is crucial to consider geopolitical factors, as these materials are becoming a part of their national security. However, the determination of the environmental impacts of mining and processing is still ongoing as the world accelerates into the electrification movement (Figure 2.3). The topics of environmental and ethical considerations for handling these chemicals are only now surfacing, due to the sheer volume of materials being processed as production scales from the Gigawatt level to the Terawatt level.

PRACTICAL INSIGHTS | Considerations of the Global Mining Industry

Long lead times of at least seven years – and more typically 10–20 years – to bring new mining and refinement operations online mean that today's investments will impact the rate of decarbonization as far ahead as 2040. Current investments are enough to meet the demand for a gradual transition, but this would fall short of achieving the Paris Agreement objectives. Without major new investments, there will be only half the lithium required in 2030 to meet climate targets. Even with greatly increased levels of investment, without reducing the number of EVs and increasing recycling rates, we will run out of lithium.

The large variability in the time to commission new mines is predominantly driven by differences in permitting regimes. With investments and permits secured, it would be possible to bring a new mine into a functioning state in a minimum three years. Permitting depends on local attitudes to new mines. Countries such as Argentina, Australia, and Canada have established mining industries, which are located far from major population centers. Such countries require relatively short permitting times. In the United States

(continued)

> **PRACTICAL INSIGHTS** | Considerations of the Global Mining Industry (*continued*)
>
> and Europe, local opposition to mines can cause delays of up to 25 years. For national and investment stakeholders, efforts to reduce environmental impacts are focused on reducing energy use and promoting electrification. However, for the local stakeholders, water usage is a major concern, and this often becomes the key factor influencing permitting times or decisions. While the environmental impacts of mining are still relatively unknown, runoff leading to heavy metal contamination of groundwater is not considered a major risk. Depletion of aquifers may be a more significant consequence, forcing the industry to shift toward reducing water consumption.
>
> Lithium is typically extracted from either brine-containing lithium chloride or hard rock mines containing lithium oxide. NMC and NCA batteries generally require lithium hydroxide, whereas LFP uses lithium carbonate. Most of the world's lithium comes from five hard rock mines (i.e. four in Australia and one in China) and six brine operations (i.e. two each in Argentina, Chile, and China). The 2.5 million tons of lithium metal, or 13.3 million tons of lithium carbonate equivalent (LCE), required to meet EV production demands, is 12 times the current global extraction rate of 1.1 million tons of LCE. At this rate, it would deplete the known reserves of 140 million tons of LCE in under ten years. The untouched and unexploited Clayton Valley lithium clay deposits in Nevada are believed to contain a total of 18.1 million tons of LCE. The world's oceans contain an estimated 954 Gt of LCE, but extracting the required quantities would lead to the processing of vast quantities of water (i.e. the current evaporative methods of extraction from lithium-rich brine would not be practical). Various methods involving sorbents, dialysis membranes, and electrodes are under development but are still far from being economically feasible.
>
> Global copper mines currently produce 21 million tons annually with known reserves of 880 million tons. The largest single use is currently in the construction of buildings. EV production and electrical grid upgrades will consume as much as 36 million tons per year, requiring a massive increase in copper production, whereas installing catenaries to electrify major roads and railways would require a relatively insignificant amount. Although this is certainly a challenge, it is somewhat mitigated by high recycling rates, meaning that while mine output must be increased, there is little risk of running out of copper. Copper mining operations are well distributed globally, further reducing the risk of major supply disruptions. Significant extraction occurs in 14 countries distributed over all 6 continents, with notable producers being Chile (14%), Peru (10%), China (8%), the Democratic Republic of the Congo (8%), the United States (6%), Australia (4%), Russia (4%), and Poland (2%). A further 14% of global production comes from a large number of countries each producing much smaller quantities. While increasing demand may lead to price spikes and temporary shortages, it is expected that copper shortages will not ultimately prevent decarbonization.
>
> Current mine production of silver is 24 kt annually from known reserves of 530 kt. Without significant reductions in the silver intensity of solar panels, their manufacture will require 45–112 kt. Although crystalline silicon cells are reducing their use of silver, expanded mining operations will be required unless perovskite cells disrupt this market.
>
> The shift to domestically produced renewable energy from imported fossil fuels will improve energy security and insulate economies from spikes in fuel prices. However, geopolitical vulnerabilities will not go away. The supply of the materials required for a decarbonized energy system often depends on a very small number of countries. For example, China currently extracts 71% of the world's silicon and produces more than 80% of the world's polysilicon.
>
> Source: Muelaner, J., "Critical Metals, Sourcing, and Long Supply Chains: Constraints on Transport Decarbonization," *SAE Research Report* EPR2022SE2, 2022, **https://doi.org/10.4271/EPR2022SE2**.

Why Source Selection Is Important

Lithium is essential to the batteries being deployed in transportation batteries today. These batteries vary in many ways such as the type of other metals used in the cathode, including shape (cylindrical, prismatic, and pouch being the typical formats) and size of a battery. Lithium is an essential element in these batteries due to its inherent nature of having the highest electrochemical potential of any metal, which allows for greater energy storage in a comparable size.

Lithium Discovered

Lithium is one of the more common elements in the earth's crust ranking 32nd and standing among common elements such as aluminum, sodium, and potassium and of the same order of magnitude as nickel, cobalt, copper, and zinc. Johan August Arfvedson discovered lithium in Sweden in 1817, and with similar properties as sodium, potassium, cesium, rubidium, and francium, it was classified in the alkali metal group. In nature, it is not found as an isolated element but rather in the form of a compound. The two main sources are as follows: first, pegmatites that are formed from granitic magmas and typically are a combination of lithium, aluminum, and silica depending on the mineral type; the second source is subterranean brine reservoirs that are the result of surface water eroding volcanic sources into this closed reservoir. In a brine resource, lithium is found in the form of lithium chloride with other elements present depending on the nature of the surrounding environment.

Early Uses

Lithium found in its early uses works as a compound in alloyed components to reduce wear in friction on railway axles. But the first significant growth period for lithium in industry was for the production of lithium deuteride which enabled thermonuclear fusion and the first atomic weapons. Today, this is still the case but lithium is used more as the initiator for the fission of natural uranium than as the source itself. After this growth in demand for lithium, a number of other applications began to surface. Those applications include the use of lithium in both glass and aluminum production as a flux material, as a thermal stability aid in glass, and as a constituent in stearate-type greases. In the late 1970s, researchers developed the first lithium batteries where lithium metal was used as anode material. Further research led to the development of a more stable battery where lithium was mixed with other metals to form the cathode side and graphite was used as the anode. Three prominent members of the industry were awarded the Nobel Prize in Chemistry for their contributions to the developments that led to the commercialization of today's LIB: they were Drs. Stanley Whittingham, John Goodenough, and Akira Yoshino. These developments allowed for batteries that had superior energy density compared to other options (lead acid, nickel metal hydride, and nickel-cadmium) which in turn set a pathway to expanded portability for electric consumer items such as music players and cellphones. The commercialization of LIBs began in the early 1990s with products such as Sony's Handycam video recorder. By the late 2000s, LIBs were being deployed in EVs such as the Tesla Roadster, Nissan Leaf, Chevrolet Volt, and Mitsubishi i-Miev and led batteries to account for nearly one-third of lithium demand (Figure 2.4).

34 CHAPTER 2 Sourcing Lithium-Ion Batteries

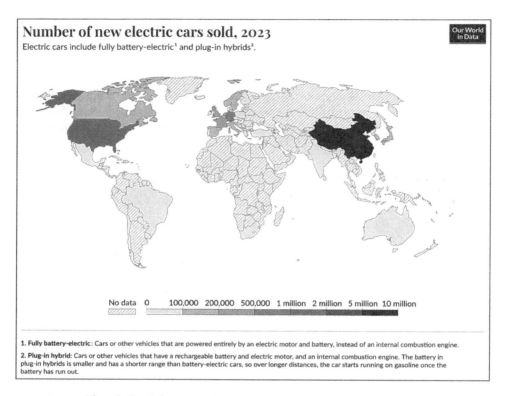

FIGURE 2.4 Electric Car Sales. *Source: https://ourworldindata.org/grapher/electric-car-sales/CC BY 4.0/ Our World In Data.*

This 2023 illustration, maps where EV sales have proliferated. The Chinese government has made large-scale investments in the development of an EV industry and an associated battery manufacturing industry. Some industry experts have assessed the Chinese as technologically more advanced than the rest of the world in battery development for several years.

PRACTICAL INSIGHTS | M. Stanley Whittingham and the Evolution of the LIB

M. Stanley Whittingham conducted much of the early development work on the theory of intercalation, where lithium atoms are inserted into a material (intercalation) and then returned to their host material in the same form. This process results in a reversible reaction (discharging–charging). John Goodenough conducted early research at the University of Oxford in the United Kingdom developing some of the earliest versions of LIB materials with a focus on LCO (Lithium Cobalt Oxide) chemistries. He later, at the University of Texas, developed the spinel class of materials, most notably LFP. Akira Yoshino was a research scientist with Asahi Kasei, and conducted further research on the work of Goodenough and combined the cathodes developed previously with graphite materials. His work was commercialized by Sony and ultimately resulted in the LIBs used today.

> **PRACTICAL INSIGHTS** | M. Stanley Whittingham and the Evolution of the LIB (*continued*)
>
>
>
> Nobel Laureate M. Stanley Whittingham, widely recognized as the "father" of the lithium-ion battery. *Source: AmeliaHawneyBenchmark Wikimedia Commons/CC BY-SA 4.0.*

Today, batteries account for more than 80% of lithium demand and have proliferated into applications including grid-based storage, and unmanned aerial vehicles along with the highest profile application of EVs. Of the major constituents of these batteries, lithium garners the most attention because not only it is integral to all of these batteries but also it is responsible for significant part of the cost. Other elements in these batteries include Nickel, Cobalt, Aluminum, Manganese, Iron, and Phosphorous (Nasajpour-Esfahani et al. 2024). Nickel and Cobalt too have similar paths as Lithium because many of the dominant battery types today rely on these three elements and their sourcing needs. Sourcing requirements for these three elements have been under the spotlight as consumers have focused on the impact of their actions and the comparison with the historical products and their sourcing.

Where Do These Products Come from?

All of these products require mining, as do legacy propulsion methods. As an example, the iron and aluminum in internal combustion engines are sourced from mines all over the world with the largest sources found in Australia while a minority of the material is sourced from recycled feedstock.

Mining is a subject that instills very strong feelings in most discussions. Many people see mining through the lens of historical mineral extraction where vast quantities of land are disturbed and tailings dams are utilized. In reality, today's mining methods are very sophisticated, disturbing as little land as possible (lowering the cost since as many mining engineers will tell you every kilogram moved is costly) and commitments to restore the land to as close to the previous state, as much as possible. What many fail to recognize is the idea that all extraction is mining. For example, drilling for oil or natural gas is "mining" but normally it is not perceived as mining. In an ever-evolving world, reducing our need for mining is paramount but eliminating it is not a realistic choice today. By shifting our energy consumption from fossil fuels to electricity society, we can reduce the need for mining by harnessing renewable sources of energy such as solar, hydroelectric, and wind. Additionally, this will in return create new opportunities to reduce mining since these materials in many cases can be recycled back into their original state capable of being reintroduced to the supply chain. A few authors have called this recycling "urban mining."

Case Study: Batteries Go Underground: An Executive's View of EV Mining Equipment by Xavier Iraçabal

An underground mining battery electric vehicle (BEV) is a relatively unusual application for onboard Li-ion battery systems, with demanding usage patterns requiring total reliability for five years, even when subjected to extreme charge/discharge cycling. *Source: SAE International.*

Diesel-powered mining vehicles add risk and complexity underground. Operators need extensive ventilation infrastructure to manage the exhaust gases, and this becomes exponentially larger as mines go deeper. That can be significant in terms of cost and space for today's deepest mines, which reach depths of 4 km (2.5 miles).

Around 80% of the energy demand for underground mining is associated with load, haul, and dump (LHD) machines and other vehicles that transport people, equipment, and materials. Diesel engines are a source of CO_2 and particle emissions, so switching to battery power is an opportunity to decarbonize and improve air quality underground.

Despite these advantages, operators have held back from electrifying vehicles because of past limitations on battery technology. However, today's LIB technology has matured to the point where it is economically viable and reliable.

Xavier Iraçabal, Saft mobility product manager.
Source: SAE International.

Battery Swapping or Fast Charging?

What operators want is total reliability from vehicles during five years of heavy use in the toughest environments. Batteries need to deliver high power over shifts of around four hours. They also need to support frequent charge and discharge cycling, 24-hour operation, and high ambient temperatures.

Two schools of thought have emerged as solutions to deliver this: battery-swapping and fast charging. Battery-swapping calls for two identical sets of batteries, with one installed and powering the vehicle and the other charging. At the end of a four-hour period, the vehicle will pull up to a swap-and-charge station, where the battery will be swapped out for a fresh battery. This normally takes 15 minutes or so while the driver has a break.

This strategy uses slow charging that does not place a significant burden on the mine's existing electrical infrastructure. However, the battery changeover adds an additional task for the underground team and requires lifting equipment for safe handling.

The other approach uses fast-charging batteries that are permanently fitted into the body of the vehicle. The latest Li-ion electrochemistry is amenable to fast charging within around 10 minutes during breaks and shift changeovers, as well as "opportunity charging" during brief pauses in normal operation when the vehicle is waiting.

Saft's battery system is a standardized 48-V building block that incorporates the Li-ion cells together with voltage and temperature monitoring systems.

(continued)

Case Study: Batteries Go Underground: An Executive's View of EV Mining Equipment by Xavier Iraçabal (*Continued*)

This approach requires specialized charging stations with a high-power grid connection. Therefore, operators may need to upgrade the mine's electrical infrastructure or install wayside energy storage, especially for large fleets that may need to charge simultaneously.

Choosing the right Li-ion blend

LIBs are familiar to consumer devices and electric cars, but Li-ion is an umbrella term that covers a wide range of electro-chemistries. These can be used singly or blended together to precisely deliver the right properties across a range of five factors: energy density, cycle life, calendar life, fast charging, and safety.

The most common Li-ion types have a positive electrode made from lithium NMC, lithium manganese oxide (LMO), and LFP. Their negative electrode is typically graphite or another form of carbon. Of these, NMC and LFP are most commonly used in underground mining vehicles, particularly the swappable types. Both provide long autonomous operation with charging taking less than one hour.

To compare the two at the level of individual battery cells, LFP has a lower voltage. When scaled up to a large EV battery system, this means that LFP needs around 30% more cells than NMC for the same system voltage and energy. However, the raw materials in LFP batteries are more abundantly available, giving them price stability for mine operators.

A new addition to the Li-ion family is lithium titanate oxide (LTO). It is similar to NMC but instead of graphite, the negative electrode is based on LTO. This gives it the ability to accept very high charging power, making charging times as short as 10 minutes. Another benefit of LTO is its ability to withstand three to five times more charge and discharge cycles than other LIBs. This makes it particularly well suited to underground mining, as does its inherent safety, even when subjected to electrical abuse such as short circuit, mechanical damage, and even deep discharge to zero volts.

A drawback of LTO is that its energy density is lower than both LFP and NMC. However, ultrafast charging means the batteries remain in place inside the vehicle, so they do not need to be mounted in an accessible location for swapping.

Another important factor for mine operators is the battery management system (BMS), which constantly monitors the voltage and temperature of cells and manages charge and discharge to keep temperature constant across the entire system. This maximizes the lifetime and ensures consistent performance. The BMS also monitors the state of charge (SOC) and state of health (SOH). Both are important measures for operation and maintenance.

Plug-and-Play Modules

Vehicle designers face practical challenges when considering battery power. Large underground mining vehicles typically need a voltage of 650–850 V. That is likely to remain the case for the foreseeable future as it avoids the increased systems costs for higher voltages. At these voltages, designers will need systems with an energy storage capacity of 150–250 kWh. Some manufacturers are looking for 300 kWh or higher, depending on the vehicle.

Vehicle manufacturers also typically want to use a modular approach. The logic is that they can design one electrical system as a basic framework. They can then apply it to multiple vehicles in their portfolio. This minimizes the amount of development time and type testing required for each vehicle, helping to keep costs and time to market under control. Recognizing this, Saft has created a battery system as a standardized 48-V building block that incorporates the Li-ion cells together with voltage and temperature monitoring systems. The system will be available in NMC and LTO electrochemistries, providing a basis for vehicle designers to build bespoke systems.

A fixed LTO battery rated at 800 V with 250-Ah capacity would deliver 3 hours of operation with a 15-minute break for fast charging. *Source: agnormark / Adobe Stock Photos.*

Inside the modules, Saft is using cells that are prismatic in shape, rather than the traditional cylindrical shape. Because prismatic cells are rectangular in cross section, there are no gaps between cells, allowing maximum energy density and battery performance to be packed into the available space.

The battery system also can be complemented with components such as a heavy-duty enclosure, thermal management system, or fire suppression system, depending on the OEM's needs. In addition, operators can monitor the performance of the battery systems across their entire fleet of vehicles.

Evaluating LHD Scenarios

As a practical comparison, Saft evaluated two alternative scenarios for an underground LHD vehicle using swappable and fast-charging batteries. In both cases, a LHD machine was considered that weighs 45 tons unladen and 60 tons fully loaded with 6–8 m3 of material. The evaluation also envisaged vehicles with batteries of similar weight and volume of 3.5 tons and 4 m3 in an envelope measuring $2 \times 2 \times 1$ m. This enables a like-for-like comparison of the two approaches.

For battery swapping, the battery system was based on NMC or LFP chemistry. It would support six-hour shifts for the LHD vehicle from a battery rated

(continued)

Case Study: Batteries Go Underground: An Executive's View of EV Mining Equipment by Xavier Iraçabal (*Continued*)

at 650 V and with 400-Ah capacity. When swapped off the vehicle, they would require three-hour charging. They would deliver 2,500 cycles over a calendar life of three to five years.

When applied to fast charging, a fixed LTO battery of the same size and weight would be rated at 800 V with 250-Ah capacity. It would deliver three hours of operation with a 15-minute break for charging. The LTO battery would last 20,000 cycles, giving it a calendar life of five to seven years.

NMC and LFP are the most commonly used chemistries in underground mining vehicles, particularly for slow charging with batteries swapped between shifts.
Source: Luca Flor/Adobe Stock Photos.

While these scenarios provide an interesting comparison, a real-world vehicle designer can apply the modular approach to develop a battery system that will precisely suit the preferences of their customers and the needs of individual vehicles. For example, they could extend the duration of the shift by increasing the size of the batteries.

Ultimately, mine operators will choose the vehicles that are best suited to the configuration of their mines. That will require a fine balance to consider the available space underground, as well as the accessibility of high-power electricity. The TCO of the vehicles and the infrastructure needed to support them also will influence decisions. That is why it's important for battery companies to provide vehicle designers with flexibility.

Source: Iraçabal, X., "Batteries go underground," Mobility Engineering, June 1, 2021.

Future Sources of Supply for Lithium

When evaluating sources of supply, particularly in Lithium, a buyer can be faced with a number of sources that can be in some ways bewildering. Much of the challenge can be attributed to the lack of information that exists in the market. As demand is set to grow rapidly in the latter half of the 2020s much of the new supply is not in operation today. In fact, nearly two-thirds of the new supply is targeted to come from new companies that are not operating companies and in many cases unfunded and are typically presenting optimistic scenarios of their operations in order to encourage investors to support their efforts.

An example of this can be evidenced in the DLE supply market. DLE stands for "Direct Lithium Extraction" and has been widely suggested as an alternative to traditional brine harvesting of underground closed basin aquifers where the majority of lithium brines are sourced. The use of DLE methods allows the development of brine sources that typically have not been economically viable due to lower concentrations compared to existing terrestrial mining operations. This is a result of selectively removing the lithium ion from large volumes of brine without the need to increase the concentration using evaporation whether it be solar or thermal. Today, the two largest brine operators are Albemarle and SQM who operate evaporative brine operations in the north of Chile in the Salar de Atacama where the underground brine pools contain approximately 2,000 ppm of lithium. The two operations in Argentina owned by Arcadium have concentrations of lithium ranging from just below 500 ppm of lithium to above 700 ppm (Figure 2.5). DLE's promise is to capitalize on large brine fields with lower concentrations in some cases down to 25 ppm. In the DLE market, it is proposed that an underground brine would be brought to the surface, allowing the lithium to be removed from the brine and then reinjecting the spent brine to the aquifer.

Rarely mentioned are the technical challenges associated with these techniques. First, many require some form of reagent to extract the lithium from the brine. Today,

FIGURE 2.5 Salt and lithium are extracted from this mine in the Salinas Grandes Salt Flats in Argentina. In the early 2020s, the search for greater supplies of Lithium focused in this region. *Source: Henrik Dolle/Adobe Stock Photos.*

these are typically classified into four types of methods: Alumina adsorption, ion exchange resin, membrane filtration, and solvent extraction.

- For Alumina adsorption methods large quantities of fresh water are normally used to strip the Lithium ion from the host alumina adsorbent material and create a lithium-rich material that is then reacted with Sodium Carbonate to form Lithium Carbonate.
- In the ion exchange method, the lithium ion is stripped from the host resin with hydrochloric acid forming a lithium-rich material suitable for conversion into other lithium compounds.
- The third type of DLE is the use of a membrane material to allow the migration of lithium ions from one solution to a virgin solution that can then be converted into a lithium compound. In this method, energy is used to force the lithium ions through the membrane under pressure.
- The final method uses a solvent (typically an organic material) that absorbs the lithium ion and then is washed with a water solution to produce a concentrated lithium solution.

All of these methods classified in the DLE category require further development to achieve commercial success and, in many cases, must be adapted to individual sources of brine further complicating the commercialization process. The promise that they bring though is the opportunity to tap into sources that previously have been unattractive either due to cost or sustainability. This would be a notable accomplishment for the industry and reduce the impact of other forms of mining as well as potentially lowering the cost of production which would likely reduce the cost of lithium and batteries.

Existing Sources of Lithium Supply

The two legacy sources of supply involve either extracting lithium from closed basin brine aquifers as mentioned above or hard rock mining. In the case of closed basins, brine is pumped from subterranean aquifers typically anywhere from as shallow as 10 m to as deep as 100 m. This brine is then put through a series of evaporation ponds where less soluble salts such as potassium, sodium, and magnesium salts precipitate out of the solution. These salts are routinely harvested from the pond system and either sold or stockpiled. The final solution becomes a lithium-rich brine that can then be chemically treated and then converted into a lithium product suitable for commerce. In the case of hard rock sources, lithium-bearing minerals are bound in the rock material. These minerals are concentrated using physical methods and then chemically extracted from the material. This can involve the use of temperature, pressure, or aggressive chemical reagents such as sulfuric acid. Typically, the extraction of lithium from hard rock sources is more capital-intensive since two different types of plants are required. The first is the concentration step involving physical methods to increase the concentration of material from levels ranging from 0.5% lithium content (5,000 ppm) up to 1.3%. This material is concentrated up to approximately 2.5% lithium content. This material is transferred to a second plant where the lithium is extracted from the mineral in a chemical conversion plant. Depending on the desired final product further chemical steps may be employed for example converting from either lithium sulfate or lithium carbonate to lithium hydroxide.

In summary, the concentration of lithium in the resource will have a very outsized impact on the cost of production and as the industry grows larger and seeks out lower-quality sources of supply, the marginal cost of production will rise, increasing the cost of lithium materials but also make it critical to seek out lower cost and stable sources of supply to avoid supply disruptions.

Sourcing of Other Minerals

Nickel

For battery-grade Nickel, much of the current supply is from the Pacific Region with the largest source being Indonesia based and accounting for exactly 50% of the global production in 2023. The Philippines and New Caledonia make up the second and third largest sources and the three together account for roughly 2/3 of the global supply (National Minerals Information Center n.d.). The largest application for Nickel is for alloying to produce stainless steels. Much of the ore is laterite where the Nickel is found with Iron deposits (Keskinkilic 2019).

Production from these sources typically has extremely high-energy consumption resulting in considerable carbon footprints. When considering different suppliers of Nickel precursor material, it is important to evaluate them on the source of minerals and how and where it was processed to gauge the potential for carbon emissions within the process.

Cobalt

The majority of cobalt today (74%) is sourced from the Democratic Republic of Congo (DRC) (Gulley 2022). Ensuring that management practices and standards are in line with the desires of Western automakers is a considerable challenge. Much of the material sourced there is shipped to China for processing into precursor materials for battery applications. According to Benchmark Minerals, nearly 76% of cobalt refining occurs in China. This dependence on single nations for the mining and refining not subject to Western standards of production creates a challenge for sourcing these desired minerals and refining them into battery-grade precursors.

In the DRC, it is common for some of the minerals to be sourced from "artisanal" mining where workers hand mine them rather than using large equipment and processes for mining. This type of mining can expose considerable hazards to the workers and in some cases may include child labor. These paths from mineral to refining create a risk to the supply chain that end users should investigate thoroughly to ensure that they are not supporting activities that do not meet the expectations of consumers.

Deep Sea Harvesting

A recent development in the industry is the creation of new opportunities for retrieving the rich minerals of the deep oceans in the form of "nodules," which are formed by a combination of many elements of the periodic table. These elements are thought to come from underwater volcanic plumes which emit elements that are dissolved as salts into the seawater. These nodules have higher concentrations of the elements such as Copper, Nickel, Cobalt, Manganese, and a variety of rare earth elements, than terrestrial-based mines. These salts are then combined into these nodules by unknown processes. However, the leading thoughts of these nodule formations are from bacterial digestion or simple electrochemical deposition. The extreme depths of the ocean where these nodules form make it extraordinarily difficult to know if these hypotheses are correct or not, but one thing is certain: it takes an extremely long time to form these nodules.

The two most prevalent methods of retrieving these nodules are deep-sea dredging and selective harvesting. The former sucks up the nodules along with a few centimeters of the ocean bed floor and is not selective with what it picks up. The nodules and silt are lifted up a riser, which dumps the silt back into the ocean creating a large plum of light-blocking materials. The concern here is the effects on the phytoplankton in the ocean. The latter utilizes a form of selective harvesting which uses a video system equipped with an artificial intelligence interface to identify life forms on nodules down to 1 mm in size and avoid those nodules with life forms. Additionally, this method uses a robot arm extending down to the ocean bed floor avoiding significant silt bed disturbances in which microscopic organisms preservation is maximized. These systems can also be programmed to only select whatever percentage of harvestable nodules without lifeforms can be picked up. This could mean leaving 20–50% of harvestable nodules behind for the preservation of the ocean bed floor with minimal disturbance. According to some, this approach is far better than terrestrial-based mining processes which have a huge environmental impact (Impossible Metals n.d.).

Cost vs. Price – The Promise of New Entrants

Many new players entering the supply of products to any market almost always use the promise that they will be lower cost than incumbent sources of supply. Discerning whether a potential new supplier has the opportunity to change the cost requires an end user to develop a rather deep set of skills to assess these claims of lowering costs. This may involve geologists to evaluate the supplier's resource potential, process experts to validate carbon consumption claims, and specialists to understand the quality impacts of these new supplies. In most cases, new sources of supply would typically be found on the right hand of any cost curve representing third- and fourth-quartile costs when using these lower-grade and existing production routes. Today, many of the new sources are proposing to utilize new technologies that would radically impact the cost of production. Examples include the before-mentioned DLE technologies in lithium to higher-grade nickel sources that previously were not utilized due to other impacts and or locations.

Contract Length and Leverage

The shift to sourcing the materials required for these new drivetrains will move the focus from the cost of a source to both the security of supply and the stability of price over a period of time. Price will still be important but the idea that both buyer and seller need some level of certainty to secure the investments they are making on both ends of the supply chain will likely result in contracts considerably longer than what has historically been seen and likely approaching 10 years. The leverage in negotiations will definitely shift from buyer to seller as very few options will be available to create a competitive landscape of suppliers.

Capital Support

One of the biggest challenges faced by new entrants is the ability to source the capital needed to build out these facilities. One area that these new entrants have pursued in working with end users is the concept of utilizing agreements either as a sign of endorsement allowing the new entrant to raise funds from outside nonaffiliated sources such as government grants and loans, as well as commercial sources of funding. Additionally, there have been prepaid agreements and lending facilities where end users have put forth capital to be used in preproduction needs including exploration, development, construction, commissioning, and working capital. Potential end users should be prepared that the need for capital is going to be a part of sourcing efforts with these new entrants.

Quality and Sourcing

A key consideration for the battery supply chain is the different approaches to quality management. When starting from the end user working backward to the source material you will find a vastly different approach to how quality is instituted. At the automotive level, quality is held in high regard due to the potential risks associated with failure and financial costs. Failures in the automotive sector not only can increase costs in the production phase but can also escalate to injury and exposure to costs from either repairs or brand impact. In many cases, automakers intend to impose their level of quality management on all suppliers upstream of their operations. This burden requires a different mindset and costs. In the midstream activities quality is approached under a similar mindset but with an outcome focused on efficiency and as a tool for lowering costs. Failures in the midstream sector typically result in increased costs related to rework or inefficient manufacturing. During the mining phase of the supply chain, consistency is a much lower priority as the mining phase is designed around dealing with inconsistent feedstocks and converting those into a profitable product.

Much of the concept of automakers moving to directly purchase battery raw materials is a reversal of the effort by many automakers to outsource their supply chains. In practice, today the auto industry relies on Tier 1 suppliers (those suppliers who directly deliver to an auto plant) to manage the activities of their supply and so on all the way up to the minerals used in a vehicle. In the battery supply chain, much of this is being upended by automakers directly negotiating supply from minerals suppliers and dictating how that supply will enter their supply chain and at what tier. The quality management systems implemented over the last few decades in the automotive supply chains have developed in an effort to reduce the cost of ownership, increase the safety of vehicles, and establish a multilevel commitment in the supply chain among the tiers of suppliers. The transition to electrified drivetrains will result in a transition from a mainly internal supply chain (engines and transmissions) to an external supply chain (cell production and electric motors) although some may be joint ventures. This transition will require suppliers who are new to the automotive industry to accept quality requirements that have been installed in the automotive sector since the early 1980s. In many ways, this will require a fair amount of development in the sector since many of these companies are not as experienced in the requirements of the automotive industry including part submission warranties (PSW), production part approval process (PPAP), management of chain (MOC), and 4M (man, machine, method, and material). The mining and refining segments have never been exposed to this level of involvement from their customer base and have historically ended their commitment to engaging on quality with customers to an agreed-upon specification for the material sold.

The Evolution of ESG in the Battery Supply Chain

In today's battery systems, the primary elements used are Lithium, Nickel, Cobalt, and a few others that are more widely used in other industries such as Iron. When assessing sources of supply most often the next questions after cost are the ESG parameters of the source. ESG represents the environmental, societal, and governance properties of a given supply chain. For many of these elements, the environmental impacts are usually focused on external impacts such as greenhouse gas (GHG) emissions, water consumption, and waste generation. The societal impacts normally fall into two categories: exploitation or unfair labor practices and engagement with the nearby residents of a project.

As noted above, Lithium is a rather common element in the earth's crust and the countries with the highest known resources of Lithium are in order of magnitude: Bolivia, Argentina, Chile, and Australia. Both Bolivia and Argentina have had difficulty developing their resources in the past due to both technical and geopolitical challenges. Today, Australia and Chile are the two largest sources of supply of the element with much of the extracted material in Australia as hard rock mineral which is then sent to China for the chemical conversion step. In Chile, where brines are responsible for all of the production, chemical conversion takes place in Chile and the impacts are nearly all there.

Much of the ESG discussion for Lithium centers around three concerns. The first is the GHG emissions associated with the conversion of hard rock minerals into a lithium chemical which depends on the concentration of the source material (lower concentrations require higher amounts of mining activity and corresponding energy consumption) and the carbon intensity of the electrical grid. The second impact is the use of fresh water in the operation which occurs in a washing or flotation step since many of the plants are in arid regions. The final major impact is waste generation which is usually related to hard rock extraction and the waste minerals not used to produce a final lithium product. If improperly stored or without a suitable alternate use this material can both drive costs up and create other impacts.

Nickel, which is predominantly sourced and extracted in just a few countries in the Pacific Southeast (Indonesia, New Caledonia, and the Philippines), is similar to Lithium in that the major ESG concerns revolve around GHG emissions and water

FIGURE 2.6 Wastewater being discharged into the environment. Water usage is a key metric for calculating the environmental impact of battery production, particularly associated with mining. In the first half of the 2020s, many companies in EV and EV battery manufacturing found themselves needing to measure their processes through lifecycle assessments (LCAs).
Source: Vastram/Adobe Stock Photos.

impacts (both use during process and disposal). In the overall carbon impact of batteries, Nickel typically makes up the greatest impact for the ternary cathode formulations. As users become more aware of the choices available to them sources of Nickel that have a lower GHG emission impact may become more favorably viewed. There are instances of wastewater (Figure 2.6) being directly released to sensitive oceanic regions in Indonesia which again may not meet the standards expected by both automakers and end users.

Cobalt, like the other key battery materials, has sensitivity to both GHG and water but has the added challenge that many of the key sources are located in the DRC. It has been widely reported that the DRC does little to stop artisanal mining from occurring. In some cases, this artisanal mining has included child labor in conditions not acceptable for Western supply chains.

As these materials are only found in certain regions of the world and usually not in regions where there is consumption, some level of transport is expected. This transport phase of the supply chain can be accomplished in an efficient manner if properly managed. For example, at the mine site once extracted these ores are usually concentrated up to a suitable level to allow stability and minimize shipping costs. Once the final material is closer to being introduced into the end product, efforts should be made to reduce storage time and transport due to potential quality impacts including packaging failure, spills, and contamination.

When the battery markets develop and the quantity of material reaching the end of life is larger recycling and recovering the materials will become a valuable source of material reducing the impacts of the extraction phase of the life cycle. Today, the expected life of many batteries is at least ten years and the market is still in a high growth phase leading to a small percentage of material being available from recycled sources. By the mid-2030s, it is expected that recycling could easily achieve a target of recovering as much as 10% of the demand in that period. Today, the recycling of automotive batteries is still in its infancy and relies almost exclusively on manufacturing scrap. This "scrap" is collected at multiple steps in the supply chain and recycled including at the cathode maker, the cell maker, and at the automaker level. Key considerations in understanding the recycling step will be to achieve recovery rates that match the impacts of the mining activities.

After you have extracted and refined the individual elements into products that can be transported they must be transformed into materials ready for the cell manufacturing process. For cathodes in a battery, these are typically combined into a mixed metal oxide (containing forms of lithium, cobalt, nickel, manganese, iron, aluminum, and phosphorous). This step mixes the corresponding feedstocks and under heat transforms them into the cathode material. Depending on the cathode material, the amount of energy inputted can be significant. On the opposite side of the cell, the anode in today's lithium-ion cells has a majority of graphite with other additives. The graphite usually is derived from natural sources where it is mined and then processed into a material suitable for use in a cell (particle size being the key parameter). Alternately, battery-grade graphite can be formed synthetically when sourced from petroleum products. Today, nearly all the graphite used in batteries is processed into the battery grade form in China creating challenges for battery makers looking to have more influence on their supply chains.

Technological Innovations in Sourcing and Supply Chain Management

There are many technology developments in the sourcing and supply chain management field as it begins to mature, particularly in North America and Europe. China has already created a mature supply chain as it dominates the world today as a testament to its success (Jin et al. 2023). However, other countries around the world now aspire to create their own network via sophisticated digital tools for supply chain management. Artificial intelligence is now being used for machine learning on how to predict material shortages will occur and computers with sophisticated software will play a dominant role in traceability and management of these critical materials.

One of the biggest movements in the global market today is blockchain for traceability and transparency. One might ask, why? In some cases, the finger points to labor, while in others to geopolitical issues. Due to the concerns of unethical forced labor or child labor in some regions of the world, corporations, as well as the general public, want to avoid materials from companies that use these products coming from such deplorable acts. On the geopolitical front, some nations have been designated as FEOC's or "foreign entities of concern," making them forbidden to be included in some nations' supply chain networks. For example, the United States of America has issued FEOC's titles to China, Russia, Iran, and North Korea.

On the technology front, new advanced extraction methods are being developed to support more sustainable mining and material processing. This includes methods to improve yield and reduce environmental impact such as water usage in processing materials and air pollution from internal combustion engines used in mining.

Business strategies and market positioning are very important contributing factors to successful supply chain management. The need for intelligent go-to-market roadmaps is essential to success. Competitive market analysis is also key to a successful business in battery supply chains.

Supply Chain Complexity and Challenges

With any new emerging technology, the build-out of the supply chain comes with many challenges. One of the most ominous is the rapid market growth and the unpredictability of market requirements. The concern here is to undercapitalize or overcapitalize for the percentage a company hopes to achieve in a complicated rapid growth market could mean losses that will sink a company, or profits that one needs a wheelbarrow to go to the bank. The key drivers present are the increasing adoption of EVs and the aging grid network worldwide. Both of these are driving variations in predictions of market needs in total GWh of production of batteries, creating unknowns with crippling effects if interpreted wrong or wealth which makes investors proud.

Another key challenge is the supply chain bottlenecks. These could be as simple as the inadequate supply of raw materials from mining due to long timelines to open mines due to the lack of permits to mine and a productive process to refine the materials into the precursors necessary to make battery materials.

China's dominance in refining raw materials came from a couple of decades of modifications to the value chain feeding the ever-growing consumer electronics industry. The key point here was the whole world sent their consumer electronics requirement to China. Since these products needed batteries, China built an industry around processing raw materials to feed the hungry industry to feed the world the consumer products nations were hungry for. Therefore, the importance of refining capacity cannot be understated, as it is critical to the success of building large volumes of batteries.

Then there is the topic of off-take agreements and their importance to the supply chain. These battery-grade processed materials are typically large volumes and expensive. Companies do their best to get offtake agreements before going into production to ensure revenues achieve adequate balance sheets for investors. This is not an easy balance, but quite necessary to achieve an effective business case.

Supply chains are very dependent upon transportation and logistics. Domestic businesses in the battery field understand the need for "localization" to achieve a cost-effective supply chain to shorten the transportation distance, and improve just-in-time deliveries, with a net result of lower costs. International business is wrought with concerns of material flow disruption due to unforeseen circumstances such as weather, strikes, equipment failures, and import and export duty officers holding up shipments, among other potential issues.

Building an optimized supply chain for batteries is extremely difficult due to the large variety of materials and form factors of materials used in their construction. Strategic planning for a supply chain is necessary for success. As mentioned previously, there are typically over 75 materials on the bill of materials for a cell and many more for building modules and packs. Each of these materials has improvements in technologies to create them, production efficiencies to increase throughput, and reductions in energy utilized, with the inevitable consequence of price point volatility.

Regulations and Standards for Ethical Sourcing

Supply chains in the battery world are highly impacted by the regulatory landscape at both domestic and international levels. Key regulatory frameworks (e.g. EU Battery Directive, US SEC conflict mineral rules) can make sourcing quite difficult, however, necessary. Many of the chemicals used in making batteries can be dangerous to humans and require scrutiny in practices of handling and transport. These special needs drive a variety of industry standards and certifications. Some examples of these include international treaties and environmental standards.

Due to the large volumes of materials and their cost in this era of GWh transitioning into TWh production volumes, it is important to have properly secured digital or blockchain platforms for the sake of tracking these materials. The value of these materials contributes to the GDP of nations producing them, thus requiring monitoring for taxation, duties, pricing, avoidance of black markets, and theft.

As mentioned earlier in this chapter, the labor and human resources practices in some countries are abhorrent, creating a necessity to flag materials coming from nations with these practices. This, in turn, creates a need for transparency for ESG (environmental, social, and governance) compliance.

Many nationals around the world are now bonding together to create standards for ethical sourcing and material traceability to ensure materials are sourced responsibly to reduce the irresponsible acts by a handful of countries using child or forced labor practices. Additionally, the need for monitoring these materials for potential environmental contamination from accidental or intentional dumping can be assessed with proper traceability of these products.

> ### Future Considerations
>
> The following are challenges and solutions to consider in the question of sourcing battery materials.
>
> The rise of electrified drivetrains will force many in the industry to take a new approach to dealing with suppliers. This may include having to provide financial support across supply chains where an automaker provides a loan to a minerals supplier or locating quality teams at upstream suppliers. The same can be said about the rapidly growing energy storage market where large utility corporations are now engaged with battery manufacturers to support their growing demand. Below are some of the key challenges battery manufacturers face in the coming years, followed by some suggestions for solutions to those challenges.
>
> Challenges:
>
> - Commodity price fluctuations and long-term supply contracts.
> - Addressing the societal concerns of the ethics of work practices in procuring these critical materials involving child labor, forced labor, and exposure to toxins during the procuring and processing of the minerals.
> - Supply chain risks of pricing, availability, transportation, and utilities.
> - Identifying the impact of trade policies, tariffs, and regulations.
> - Comprehend the impact of global pandemics, wars, and natural disasters.

- New technological advancements of alternative battery chemistry and packaging making current products obsolete.

Solutions:

- New technologies for more effective refining of minerals into useful battery precursors.
- Battery chemistry innovations to reduce critical material dependency.
- Improved long-term strategies for sourcing critical materials, including mining or harvesting and processing to achieve better ESG numbers.
- A call to action by governments, corporations, and stakeholders in the battery industry. An example of this is the Li-Bridge initiative which created a list of 26 action items provided to the U.S. Government for consideration in policy, funding, and national security interests (LiBridge 2023).
- Li-Bridge Industry Report | Argonne National Laboratory (**http://anl.gov**). A path forward for a more sustainable, resilient LIB supply chain meeting ESG objectives.
- Utilizing recycling as a continued source of these critical minerals.

Words to Know

Pegmatites Pegmatites are extreme igneous rocks that form during the final stage of magma's crystallization. They are extreme because they contain exceptionally large crystals and they sometimes contain minerals that are rarely found in other types of rocks.

To be called a "pegmatite," a rock should be composed almost entirely of crystals that are at least 1 cm in diameter. The name "pegmatite" has nothing to do with the mineral composition of the rock.

Laterite Laterite is a soil layer that is rich in iron oxide and derived from a wide variety of rocks weathering under strongly oxidizing and leaching conditions. It forms in tropical and subtropical regions where the climate is humid. Lateritic deposits are a type of weathering product that forms in tropical and subtropical regions through the process of laterization. Laterization involves the leaching of silica and other soluble materials from rocks, leaving behind a residual concentration of iron and aluminum oxides.

Intercalation It is the reversible act of inserting a molecule or ion into layered materials. In the case of a battery, it is the insertion of an ion that carries a charge.

ESG A set of criteria utilized to compare sources and describe focus areas.

- **Environmental** Measuring the environmental impact from both a short- and long-term perspective. This would include toxic potential, carbon emissions, and other items that might have an impact on a project.
- **Social** How does a project impact the community and the broader society? Are fair labor standards employed, equality, and other impacts?
- **Governance** Does a project operate in a stable and transparent manner?

Cathode   The portion of a battery identified by the negative symbol. In the case of LIBs, cathode is where the ion that transfers back and forth between charging and discharging is introduced to the battery.

Anode   The portion of a battery identified by the positive symbol. Is the recipient of ions in the charging cycle.

GHG – Greenhouse gas   Typically refers to the amount of carbon emitted in production and transport.

Reagent   A substance or compound that facilitates a chemical reaction.

FEOC   Foreign Entity of Concern as defined by the US government.

Further Reading

One particularly useful survey of the complexities of battery mineral sourcing is Ernest Scheyder's The War Below: Lithium, Copper and the Global Battle to Power our Lives. New York: Atria/One Signal, 2024.

References

Gulley, A.L. (2022). One hundred years of cobalt production in the Democratic Republic of the Congo. *Resources Policy* 79.

He, L., Xu, W., Song, Y., et al. (2018). New Insights into the Application of Lithium-Ion Battery Materials: Selective Extraction of Lithium from Brines via a Rocking-Chair Lithium-Ion Battery System. Global Challenges, 2(2), n/a-n/a. **https://doi.org/10.1002/gch2.201700079**

Impossible Metals (n.d.). **http://impossiblemetals.com**

Jin, P., Wang, S., Meng, Z., and Chen, B. (2023). China's lithium supply chains: network evolution and resilience assessment. *Resources Policy* Part B.

Keskinkilic E. (2019). Nickel laterite smelting processes and some examples of recent possible modifications to the conventional route. Department of Metallurgical and Materials Engineering, Atilim University Metals, 9(9), p. 974.

Li, Z., Pang, S., and Shen, X. (2024). Effects of non-subsidized industrial policies on embedding position of power lithium-ion battery manufacturers in global value chain: firm level evidence from China. *Journal of Cleaner Production* 461.

LiBridge (2023). Building a Robust and Resilient U.S. Lithium Battery Supply Chain. February 2023.

Nasajpour-Esfahani, N., Garmestani, H., Bagheritabar, M. et al. (2024). Comprehensive review of lithium-ion battery materials and development challenges. *Renewable and Sustainable Energy Reviews* 203.

National Minerals Information Center (n.d.). Statistics and information on the worldwide supply of, demand for, and flow of the mineral commodity nickel. Nickel Statistics and Information **https://www.usgs.gov/centers/nmic/nickel-statistics-and-information**

Negrete, M., Fuentes, M., Kraslawski, A. et al. (2024). Socio-environmental implications of the decarbonization of copper and lithium mining and mineral processing. *Resources Policy* 95: 105135. ISSN 0301-4207, **https://doi.org/10.1016/j.resourpol.2024.105135**

Ren, H., Mu, D., Wang, C. et al. (2024). Vulnerability to geopolitical disruptions of the global electric vehicle lithium-ion battery supply chain network. *Computers & Industrial Engineering*, Elsevier.

CHAPTER 3

Battery Design

Joern Tinnemeyer
EnerSys, Reading, PA, USA

Introduction

Developing mobile energy storage systems based on lithium-ion cell technology is a highly multidisciplinary engineering activity. The areas of knowledge necessary for complete system-level development require knowledge of electrochemical, mechanical, thermal, hardware, software, simulation, verification, and system-level background. Furthermore, advanced modeling of performance for the vehicle must be well realized as inputs to the design to allow all necessary drive attributes across the vehicle's lifetime and varying environmental conditions. These may include a simple Sunday drive in a moderate climate, a taxi in hot conditions, or a commercial vehicle in a northern country. This chapter will start by describing the fundamentals of different cell types and characteristics, followed by evaluation of performance and its impact on design. It will examine electrochemical choices in the battery management system (BMS), address mechanical design requirements, provide an in-depth review of different pack design strategies, and conclude with fire propagation countermeasures. Throughout the subsections, special attention will be given to design choices with respect to safety consequences.

What You Will Learn in This Chapter

In this chapter, we will explore battery design. The reader will learn to:

- Differentiate between different cell formats and their characteristics.
- Understand the design and performance and cost considerations associated with different types of cells.
- Distinguish the different types of cell chemistries and their applications in vehicles.
- Introduction to different battery architectures
- Understand the characteristics of battery cell composition and safety, especially in thermal runaway.

Electric Vehicle Batteries: From Sourcing to Second Life and Recycling, First Edition.
Edited by Bob Galyen and Frank Menchaca.
© 2025 John Wiley & Sons, Inc. Published 2025 by John Wiley & Sons, Inc.

Cells

Cell Formats

Lithium-ion cells come in various physical formats, each with specific characteristics that suit different applications. The three most common formats are cylindrical, prismatic, and pouch cells. Each format offers specialized features related to design, durability, and thermal management, depending on the mission profile of the vehicle (Figure 3.1).

Cylindrical Cells

Cylindrical cells are shaped like cylinders and are one of the most traditional formats. They are typically identified by their standardized sizes, such as 18,650 (18 mm in diameter and 65 mm in length) and 21,700 cells. The design includes a tight winding of the electrode and electrolyte layers, which are encased in a metal can.

Using a cylinder as a basis of the geometric design provides excellent mechanical strength and durability, making it ideal for high-energy high Nickel content cells. Furthermore, since it is a derivative of cylindrical primary cells, a high degree of automation and processes have made this cell format the lowest-cost option available. When the thickness is well controlled, the long, thin structure allows for good heat dissipation.

The clear disadvantage from a packaging perspective is the round shape does not allow for optimal packing efficiency. Furthermore, gravimetric energy density may be affected due to the amount of metal casing used. Finally, since they must be kept thin, they are typically lower in capacity than other formats. This tends to lead to more complex connections, as more cells are needed for the same capacity. It has the consequences of less sensing for voltage or temperature as a high degree of cells (2,976 cells for 50 kWh in a Tesla Model 3) are typically used, making monitoring of every cell exceedingly expensive.

Cylindrical cells have been the workhorse for the lithium-ion cell industry for decades. Early laptop computers were based on the 186,560-cell format. In addition, larger battery packs were originally constructed using this format due to availability and energy density. Globally, Tesla still uses this format today, with multiple automotive original equipment manufacturers (OEMs) following suit. Significant work is continuing to expand cell capacity via larger radii and/or length. The 4,860 cell

FIGURE 3.1 Three types of battery cell constructions, from left to right: cylindrical, prismatic, and pouch. *Source: CRBAman/Wikimedia Commons/CC BY-SA 4.0.*

has become fashionable in modern pack design. The multiple cell interconnects plus geometric inefficiencies will still be challenging.

Prismatic Cells
Prismatic cells are enclosed in a hard, flat-sided casing, usually made of aluminum, which offers good space utilization. They contain layers (stacked) or wound electrode (jellyroll) materials inside a box-like structure. They are designed to maximize space efficiency with their rectangular shape and are often larger than cylindrical cells.

The prismatic system is highly scalable and able to fit packs well due to its rectangular shape. This gives the possibility of high packing efficiency or volumetric efficiency. Furthermore, the cells tend to have a higher capacity than cylindrical allowing for simplified fastening and cell monitoring. Finally, the large interfacial area allows for good heat dissipation.

The disadvantage of this cell format is that from a manufacturing perspective, they are more difficult to produce than cylindrical or pouch formats. Further, care must be given when using these cells with highly active electrochemistry such as high nickel content.

Prismatic cells provided the highest degree of capacity of a single cell type. The majority of vehicle battery packs and general large-scale energy storage use this cell format. The ease of manufacturing into modules or packs has certainly added to its viability. Today, large blade cells are produced that can span almost the width of a vehicle, offering exceptional volumetric energy density. However, these cells pose significant challenges when faced with high-energy electrochemistry. As a result, significant precautions are necessary for fire propagation mitigation methodologies, which can lead to substantial cost penalties.

Pouch Cells
Pouch cells consist of layers of the electrode and electrolyte sealed in a flexible, foil-like pouch. This format is lighter than metal-encased cells and allows for versatile shape and size adaptations. The absence of a rigid outer case means that pouch cells can expand and contract slightly during charging and discharging cycles.

The cells are lighter and can be constructed in various shapes and sizes, optimizing space usage and integration into different products. Generally, they offer higher energy density due to the lack of heavy casing. They are, however, more susceptible to damage during the manufacturing process. Furthermore, they must be placed in some type of modular structure as the cells lack rigidity.

Pouch cells allow for a low-cost production alternative to prismatic cells while providing increased capacity. The cells have been integrated into both automotive and large energy storage systems with successful outcomes. Their cost/benefit is usually reduced at the module/pack level as rigidity and enhanced fire propagation mitigation technologies are added.

All three formats have been used in automotive applications to date. Arguments may be given for all three formats as to their advantages or drawbacks. In general, in vehicles today, it appears to be more of a function of experience and supply chain reliability than a pure engineering decision.

Comparison Summary
Prismatic and pouch cells are better than cylindrical in terms of space utilization within battery packs. Cylindrical cells offer higher durability and better thermal

management due to their metal casing and shape, which allows for more effective heat dissipation. Cylindrical cells are generally less expensive and easier to manufacture at scale than prismatic and pouch cells. Pouch cells often have the highest energy density but require more careful handling and sophisticated BMSs to ensure safety and longevity. Choosing the right cell format depends on the specific requirements of the application, including cost, space, weight, energy density, and thermal management needs. Each format has its niche where it performs best and advancements in technology may shift these advantages over time.

Cell Chemistry

Lithium-ion cells are not uniform in construction or characteristics, rather they may be better characterized as a family of cells. The name of a particular lithium-ion cell is derived from the substances from which it is made, such as "lithium manganese," "lithium cobalt," and "lithium iron phosphate" batteries. For most cells, the cathode contains the unique chemistries. For example, lithium manganese oxide ($LiMn_2O_4$) is the cathode material used in lithium manganese cells, whereas lithium cobalt oxide ($LiCoO_2$) is the cathode material used in lithium cobalt cells. The anode materials tend to be more conserved across different cells. Most often, synthetic or natural (graphite) is used to construct the anode.

These differences in chemistry ensure that many simple, generalized attempts to monitor and manage lithium-ion cells are less than ideal. That said, the cells function in a similar way despite these differences in chemistry. Figure 3.2 highlights this similarity in function: the lithium ions shuttle between the anode electrode and cathode electrode as the cell charges and discharges, respectively.

Cathode

There are several different cell chemistries used in automotive systems. One of the most popular is Lithium Nickel Manganese Cobalt Oxide (NMC). The NMC system offers a good balance between power and energy density for electric vehicle applications. Typically, the system is described as a percentage ratio of the main metal content in the material. For example, NMC111 describes an equal ratio of Nickel, Manganese, and Cobalt, whereas NMC811 gives an 80% content of Ni with a 10%

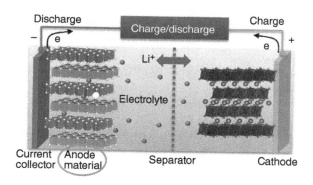

FIGURE 3.2 Lithium ion transport between anode and cathode. *Source: Chen et al. (2016)/John Wiley & Sons.*

each of Manganese and Cobalt. As the ratio of nickel increases, generally so does the energy density. Additionally, as cobalt is a more commercially volatile metal, reducing system exposure to these fluctuating costs also has a massive advantage. The concern however is that the cell becomes more energetic as the nickel content increases demanding more attention to be paid to fire mitigation techniques including restricting form factors to cylindrical cells.

Similar to NMC, but typically with even higher energy density, are Lithium Nickel Cobalt Aluminum Oxide (NCA) cathodes. In these systems, there is also a drive for this material toward higher nickel content with similar advantages and disadvantages to NMC. Moreover, this cell type is also typically more sensitive to temperature with a slightly lower cycle life compared to NMC. The cell has applications in high-performance vehicles in often cylindrical configurations.

Lithium Iron Phosphate (LFP) cathode chemistry has lower energy density when compared to NMC or NCA but offers high stability with respect to temperature. This leads to potentially different fire propagation strategies. However, care must still be taken, as the electrolyte content is higher than both Nickel-based chemistries for the same energy content. This leads to more fuel availability if the system does go in a runaway condition. Furthermore, no cobalt and nickel content gives more price reliability as some metal pricing volatility is removed. Conversely, it makes the cell type less attractive for recycling, as the intrinsic metal value is low. Due to the lower energy density, this chemistry has typically been used in larger vehicles (IE buses) where weight and space are less of a constraint. However, LFP is also aggressively entering the passenger vehicle market to reduce costs. This has mainly been enabled by cell-to-pack (CTP) constructions which will be described later in the chapter.

Anode

The following is a synopsis of the different types of anode materials used. The anode is typically described as the negative of the cell and stores lithium ions during the charging process.

Graphite (Natural or Synthetic)

Graphite is the most used anode material in lithium-ion batteries. It is favored for its stability, relatively low cost, and good energy density. Graphite anodes work by intercalating lithium ions between the layers of carbon atoms during charging. This material can store one lithium ion for every six carbon atoms, providing a good balance of energy capacity and structural stability. The graphite powder may be synthetic or natural, each having different properties.

Lithium Titanate

Lithium titanate anodes offer significantly faster charging speeds and a much higher number of charge/discharge cycles. It results in greatly extending the battery's lifespan. They operate at a higher voltage, which reduces the risk of lithium metal plating or dendrite formation,[1] thereby enhancing safety. However, LTO anodes typically have a lower energy density than graphite, which can result in lower overall battery capacity and reduced range.

---

[1] Dendrites are small filaments of lithium metal that may form under certain adverse conditions. These may puncture the separator and cause shorting.

Silicon-Based Anodes

Silicon is considered a highly promising anode material due to its ability to hold up to 10 times more per unit mass lithium than graphite. This results in potentially much higher energy densities, which can significantly increase battery capacity. However, silicon expands and contracts dramatically during charging and discharging, which can lead to rapid degradation of the anode material. To combat this, silicon is often used in combination with graphite or other materials that help mitigate the expansion and improve longevity.

Composite Anodes

Composite anodes are designed to optimize the properties of silicon while maintaining the stability of graphite via coating. These composites often mix silicon with carbon. This approach allows for an increase in the energy density (thanks to the silicon) while leveraging the mechanical stability of graphite to handle the expansion issues associated with silicon.

Lithium Metal

To target extremely high energy density, lithium metal offers the best anode material. The greatest concern is the ease with which dendrites can form. At the time of writing, this material still presents challenges and is only used in small-scale applications such as drones.

Carbon Nanotubes and Graphene

Carbon nanotubes and graphene offer exceptional conductivity and mechanical strength, which can lead to faster charging, greater capacity, and improved longevity (Kim et al. 2023). Their high surface area and excellent electrical properties help in achieving superior performance characteristics. These compounds may also be combined with standard graphite material to provide enhanced conductivity performance.

Each of these anode technologies offers distinct characteristics and the choice of anode material is driven by the specific requirements of the vehicle mission profile, energy density, charge rate, lifespan, cost, and safety. The ongoing development in anode materials is crucial for advancing cell technology and meeting the increasing demands of electric vehicles and other high-performance energy storage applications.

Electrolytes

In a cell, the electrolyte serves as the medium for lithium ions to travel from the anode (negative electrode) to the cathode (positive electrode) during discharge and back again during charging. This movement of ions is crucial for the flow of electrical current to power the vehicle.

> **PRACTICAL INSIGHTS | Advances in Sodium-Ion Electrochemistry**
>
> Cheap and abundant, sodium is a prime promising candidate for new battery technology. However, the limited performance of sodium-ion batteries has hindered their large-scale applications. In 2022, a research team from the Department of Energy's

Pacific Northwest National Laboratory (PNNL) announced it had developed a sodium-ion battery with greatly extended longevity in laboratory tests.

In batteries, electrolyte is the circulating "blood" that keeps the energy flowing. The electrolyte forms by dissolving salts in solvents, resulting in charged ions that flow between the positive and negative electrodes. Over time, the electrochemical reactions that keep the energy flowing get sluggish, and the battery can no longer recharge. In current sodium-ion battery technologies, this process happens much faster than in similar lithium-ion batteries.

The PNNL team attacked that problem by switching out the liquid solution and the type of salt flowing through it to create a wholly new electrolyte recipe. In laboratory tests, the new design proved durable, holding 90% of its cell capacity after 300 cycles at 4.2 V, which is higher than most sodium-ion batteries previously reported.

The current electrolyte recipe for sodium-ion batteries results in the protective film on the negative end (the anode) dissolving over time. This film is critical because it allows sodium ions to pass through while preserving battery life. The PNNL-designed technology works by stabilizing this protective film. The new electrolyte also generates an ultrathin protective layer on the positive pole (the cathode) that contributes to the additional stability of the entire unit.

The PNNL-developed sodium-ion technology uses a naturally fire-extinguishing solution that is also impervious to temperature changes and can operate at high voltages. One key to this feature is the ultrathin protective layer that forms on the anode. This ultrathin layer remains stable once formed, providing a long cycle life.

For now, sodium-ion technology still lags behind lithium in energy density. However, it has its own advantages, such as imperviousness to temperature changes, stability, and long cycle life, which are valuable for applications of certain light-duty electric vehicles and even grid energy storage in the future.

Source: "Designing Sodium-Ion Batteries with Extended Longevity," Mobility Engineering, September 1, 2022.

Types of Electrolytes

Liquid Electrolytes: The most common type, liquid electrolytes are typically composed of a lithium salt (such as $LiPF_6$) dissolved in a mixture of organic solvents (like ethylene carbonate and dimethyl carbonate). These electrolytes offer high ionic conductivity and are relatively easy to manufacture.

Solid electrolytes can be classified into two main categories: inorganic and polymer electrolytes. Inorganic solid electrolytes, such as lithium garnets (LLZO), offer excellent ionic conductivity and stability but are challenging to process. Polymer electrolytes, made from polymers such as polyethylene oxide (PEO) mixed with lithium salts, provide flexibility and easier manufacturing but usually exhibit lower ionic conductivity (Yang et al. 2023).

Bridging the gap between liquid and solid electrolytes, gel electrolytes consist of a liquid electrolyte immobilized within a polymer matrix. They offer a good balance of ionic conductivity and mechanical stability, making them a promising option for various applications.Research into electrolyte materials is ongoing, with several exciting developments aimed at improving battery performance, safety, and longevity. For example, ionic liquids are salts that are liquid at room temperature and have been explored as safer alternatives to traditional organic solvents. They are nonflammable and can operate at higher voltages. This not only enhances battery safety but can also improve performance.

Another group is superionic conductors. These materials exhibit exceptionally high ionic conductivities, approaching those of liquid electrolytes, but in a solid state. They hold promise for creating all-solid-state batteries with higher energy densities (Tuo et al. 2023).

The final group is a class called hybrid electrolytes. These combine the best properties of different electrolyte types, hybrid electrolytes aim to enhance ionic conductivity, mechanical strength, and thermal stability. These materials can potentially offer superior performance across various battery applications.

Separator

In a lithium-ion battery, the separator is a thin, porous membrane that physically separates the anode from the cathode. Despite its name, the separator does more than just divide these two components; it ensures that the flow of ions is regulated while preventing electrical short circuits. This balance is critical for the battery's performance and safety. There are a number of different types of separator materials discussed below.

Polyolefins, such as polyethylene (PE) and polypropylene (PP), are the most commonly used materials for separators. These polymers are chosen for their chemical stability, mechanical strength, and ability to shut down (melt) at elevated temperatures to prevent thermal runaway. Ceramic-coated separators enhance thermal stability and safety. These separators are coated with ceramic materials like alumina. This coating helps the separator maintain its integrity at higher temperatures and reduces the risk of short circuits caused by dendrite formation. Composite separators combine the properties of different materials to achieve better performance. These can include layers of polyolefin films with ceramic coatings or blends of polymers and ceramics. This, in turn, offers improved thermal and mechanical properties. Finally, nonwoven fabric separators offer high porosity and excellent electrolyte wettability. They are used in applications where high-power and quick charge/discharge capabilities are required.

Several properties are critical when selecting a separator. Importantly, the separator must have a high porosity to allow the free flow of lithium ions between the anode and cathode. The pore size must be carefully controlled to prevent dendrites from penetrating through and causing short circuits. Furthermore, they need to withstand the physical stresses during battery assembly and operation. High mechanical strength ensures that the separator does not tear or deform, maintaining its integrity throughout the battery's life. To better control fire propagation and in general enhance safety, good thermal stability is essential (Tong and Li 2024). The main attribute is to prevent the separator from melting or shrinking under high temperatures. This can lead to battery failure. Materials that can shut down at elevated temperatures are particularly valuable for safety. Finally, chemical compatibility must also be considered. The separator must be chemically inert with respect to the electrolyte and electrodes to prevent degradation over time. Compatibility ensures long-term stability and performance.

There are a few ongoing innovations to improve separator performance. For example, research is focusing on developing separators that can withstand even higher temperatures, improving safety margins. These include advanced ceramic coatings and novel polymer blends. Further innovations include separators with the ability to detect and mitigate internal shorts. These separators can incorporate materials that react to certain conditions, providing an extra layer of safety. Finally, materials such as aramid fibers (used in Kevlar) are being explored for their exceptional strength and thermal properties. These materials could offer significant improvements over traditional polyolefin-based separators.

Case Study: Solid-State Batteries

While this chapter, and much of this book, focuses on lithium-based batteries for EVs, there are other battery types emerging. One is solid-state batteries.

A solid-state battery uses solid electrodes and a solid electrolyte instead of liquid or polymer gel electrolytes like in Li-ion or lithium polymer batteries, respectively. The most commonly cited advantages of solid-state batteries, compared to their liquid counterparts, include lower flammability, increased voltage, higher cycling performance, better durability, and the ability to address instabilities between solid (electrode) and liquid (electrolyte) phases and interphase formation. Another major advantage of solid-state batteries is their increased energy density and safety. However, these advantages come at a significantly higher cost, with BIS Research claiming that the price can vary between $400 and $800 per kWh by 2026.

The above-mentioned advantages can solve many issues of traditional Li-ion batteries and be the next major automotive technology, thanks to their desired parameters. We have yet to address the durability and stability, energy density and power density, and material costs of these emerging chemistries. Additionally, lithium solid-state batteries are expected to use up to 35% more lithium metal compared to today's Li-ion batteries – an already diminishing resource. However, at the same time, they will require less graphite and cobalt. With significantly higher energy density, less cells will also be necessary for the same energy content. We will likely address at least some of these challenges as the technology matures and becomes commercial; however, tackling material costs and lithium loading will be a very difficult task.

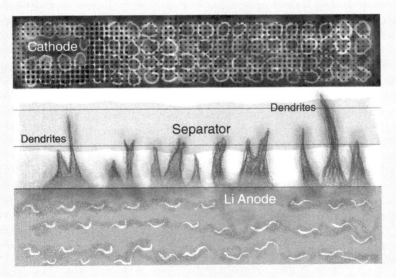

FIGURE 3.3 This drawing demonstrates lithium dendrite formation. Dendrites are tree-like structures that can form between the battery's anode and cathode. This can lead to a short circuit and other forms of breakdown in a battery's performance. Dendrites can grow in LIBs as well as in solid-state batteries.

(continued)

> **Case Study: Solid-State Batteries (*Continued*)**
>
> Metallic lithium forms dendrites on the anode side. As dendrites grow, they pierce the separator and continue growing toward the cathode (Figure 3.3). Once dendrites reach the cathode, they short-circuit the cell. Mechanisms of dendrite formation and its prevention are still a research subject. However, some possible mechanisms include dendrite formation because of micro-crack formation due to pressure induced by plating at the electrode and solid electrolyte interface (SEI), or chemical degradation of the solid electrolyte. Dendrite growth in solid-state lithium cells has been mitigated by using electrolytes that are more robust to dendrite piercing or by operating cells at elevated temperatures. Nevertheless, dendrite formation remains a serious challenge in battery technologies, and solid-state batteries tend to be more exposed to this problem.
>
> Source: Kolodziejczyk, B., "Emerging Automotive Battery Chemistries: Hedging Market Bets," / SAE International.

Cell Performance Characteristics

The following is a list of cell features that must be well understood to develop an automotive battery pack.

Nominal Capacity

The nominal capacity of a cell refers to the amount of electric charge it can store and deliver under specified conditions at its rated voltage. It is typically measured in ampere-hours (Ah). This value indicates the theoretical amount of electricity that the cell can provide at a constant discharge rate until it reaches its cut-off voltage, starting from a fully charged state.

To determine the nominal capacity, a cell is usually subjected to a standard test that involves discharging it at a constant current until it reaches the voltage threshold where it is fully discharged. This process helps in establishing a baseline for comparing the energy storage capabilities of similar cells.

Nominal capacity is a crucial metric for manufacturers and users because it provides a standard measure of a battery's energy storage potential. This is essential for sizing and designing battery systems for specific applications, like electric vehicles or energy storage systems. The capacity can be affected by various factors, including the cell's temperature, the age of the battery, the rate at which it is charged or discharged, and the operating conditions.

Energy Content

The energy content of a cell also known as the energy capacity refers to the total amount of energy that the cell can store and release. It is usually expressed in watt-hours (Wh). This metric indicates how much power a cell can deliver over a specified period making it a crucial parameter for evaluating the suitability of a battery for various applications, particularly where endurance and runtime are important.

To calculate the energy content of a lithium-ion cell, you multiply the nominal capacity (measured in ampere-hours, Ah) by the nominal voltage (V) of the cell:

$$\text{Energy Content (Wh)} = \text{Nominal Capacity (Ah)} \times \text{Nominal Voltage (V)}$$

This calculation provides a theoretical value of how much energy the cell can deliver under ideal conditions. It is important to note that the actual usable energy may be lower than the calculated energy content due to various factors such as efficiency losses during discharge, the operational environment, and the condition of the battery.

Energy Density

The energy density of a cell refers to the amount of energy that the cell can store relative to its size or weight. It is a critical measure for evaluating the efficiency and suitability of a battery especially if volume or weight are of significant concern. Energy density is expressed in two main forms:

Gravimetric Energy Density: This is the energy per unit weight of the battery and is measured in watt-hours per kilogram (Wh/kg). It indicates how much energy a battery contains in comparison to its weight, which is especially important for mobile applications where added weight can reduce performance.

Volumetric Energy Density: This is the energy per unit volume of the battery and is measured in watt-hours per liter (Wh/L). It reflects how much energy can be stored in a given volume, making it crucial for applications where space is limited.

Operating Voltage

The operating voltage refers to the voltage range across which the cell can safely and effectively discharge its stored energy. Typically, this range starts from the fully charged voltage down to the cut-off voltage where the battery should no longer be discharged to prevent damage or reduced lifespan. In most cases, working areas of the cell at set in a window between maximum charge and discharge voltage.

The nominal voltage is the average voltage that a cell maintains while discharging under normal conditions. For most cells, the nominal voltage is often around (3.6–3.7 V). Phosphate-based variants such as LFP may have a lower nominal of 3.2 V. When fully charged, a typical cell has a voltage of approximately 4.2 V. This is the maximum voltage that the cell should reach during charging. The minimum voltage level to which a cell should be discharged before charging is required is called the cut-off voltage. Discharging below this level can cause irreversible damage. The typical cut-off voltage for most cells is around (2.5–3.0 V), though this can vary based on the specific chemistry and design. The operating voltage window (from fully charged to cut-off voltage) is crucial for the safe and efficient use of the battery. It also determines the power management strategies in the devices that use these batteries. The charging electronics and load must be designed to operate efficiently within this voltage range. Moreover, maintaining operation within this voltage range helps maximize the battery's lifespan and performance by avoiding stress on the cell's materials.

Operating Temperature

The operating temperature of a cell refers to the range of temperatures within which the cell can function effectively and safely. A significant level of control and sensing is required to keep the cells within this range. If the temperature is too low during charging, dendrites may easily form presenting possible future shorts. If the temperature is too high, deformation of the separator may result in possibly thermal runaway effects.

Most cells operate best within a temperature range of about 20–25°C. This range provides optimal conditions for the chemical reactions necessary for storing and releasing energy efficiently. At temperatures below approximately 0°C, the performance of the cell starts to diminish. The internal resistance increases which can cause slower charging rates and reduced power output. The electrolyte can become more viscous, impeding the flow of lithium ions between the anode and the cathode, which reduces the cell's efficiency and capacity. If high charge currents are maintained lithium metal dendrites may start to form with the potential of leading to shorting.

Above about 45°C, cells may also begin to degrade more rapidly. The higher temperatures can accelerate the breakdown of the electrolyte and other internal components, leading to diminished capacity and shortened lifespan. Extremely high temperatures, typically around 80°C and above, can pose significant safety risks, including thermal runaway – a condition where the cell generates more heat than it can dissipate, potentially leading to fires or explosions. This is typically caused by the deformation of the separator.

In order to maintain safety and performance, lithium-ion batteries often incorporate thermal management systems. These systems can include a radiative plate with a liquid that can either cool or heat the battery depending on conditions.

Internal Resistance

The internal resistance of a lithium-ion cell refers to the opposition within the cell that impedes the flow of electric current. It is a critical parameter that affects the performance, efficiency, and heat generation of the battery. Internal resistance is influenced by various factors, including the materials used in the cell, the cell's construction, the state of charge (SoC), the age of the cell, and the operating temperature.

Internal resistance is typically measured in ohms (Ω). It can be determined by applying a small current to the battery and measuring the resulting voltage drop, or through more sophisticated methods like electrochemical impedance spectroscopy, which provides a detailed analysis of resistance at different frequencies.

Several components make up the internal resistance. These may then be combined into electrochemical models for advanced predictive analysis. Ohmic Resistance (R_{ohm}) is the resistance from the electrodes, current collectors, and electrolyte. It can be directly measured when the cell is at rest. Charge Transfer Resistance (R_{ct}) is the resistance that occurs at the interface between the electrodes, and the electrolyte and involves the resistance for the transfer of electrons. Mass transport resistance includes limitations due to the diffusion of ions in the electrolyte and through the electrode materials.

The resistance of a cell will have several effects. For example, higher internal resistance results in lower efficiency as more energy is lost to heat during charging and discharging. This loss manifests as a voltage drop under load, meaning the battery cannot deliver its full potential voltage. Furthermore, as resistance causes energy losses in the form of heat, high internal resistance can lead to significant heating of the cell, especially under high load conditions or during fast charging potentially posing safety risks. The internal resistance also determines the maximum current drawn from the cell. Higher resistance limits the peak current, reducing the power output.

Internal resistance varies with the SoC. Typically, it is lower in the middle of the cell's charge range and increases toward the ends. It also changes with temperature, decreasing as the temperature rises due to increased ionic mobility in the electrolyte.

Furthermore, as a lithium-ion battery ages, its internal resistance generally increases due to degradation mechanisms such as electrode material breakdown, loss of active material, and SEI layer growth. This aging leads to reduced performance and faster capacity loss.

Understanding and minimizing internal resistance is crucial for enhancing battery performance, extending lifespan, and ensuring safety in various applications. Manufacturers continually research and develop new materials and cell designs to reduce internal resistance and improve the overall characteristics of lithium-ion batteries.

Cycle Life
Cycle life is a critical metric for assessing the economic and environmental impact of batteries. A longer cycle life means the battery can serve more reliably over a longer period, reducing the need for frequent replacements and the associated costs and environmental impact. The cycle life of a lithium-ion cell refers to the number of complete charge and discharge cycles that the cell can undergo. The typical end of life of the cell is set at a threshold of 80% of the initial capacity. Cycle life is determined through repeated charging and discharging of the battery under controlled conditions. The cell is cycled between its upper cut-off voltage (fully charged state) and lower cut-off voltage (fully discharged state). This process continues until the cell's capacity degrades to the predefined level. Care must be considered since for electric vehicles a battery never fully cycles. There are always regenerative braking pulses that will charge the battery. Furthermore, the battery may be charged after completing a minor trip with limited change in charge level.

There are several factors influencing cycle life. For example, the depth of discharge is the extent to which the battery is discharged during each cycle significantly affecting its cycle life. Shallow discharges (low DoD) can extend the cycle life, while deep discharges (high DoD) tend to reduce it. Faster charging can stress the battery more than slower, controlled charging, potentially reducing cycle life due to increased heat and more rapid degradation of materials. Another important aspect to consider is temperature. Operating and charging the battery at high temperatures can accelerate chemical reactions that degrade the battery materials, thereby shortening cycle life. Similarly, very low temperatures can also negatively impact cycle life by increasing internal resistance and stressing the battery during charging. Lastly, the battery design and different chemistries have varying tolerances for charge and discharge cycles, impacting their overall cycle life.

In applications like electric vehicles (EVs) and renewable energy storage, where batteries are regularly cycled, the cycle life is a crucial consideration for both the reliability and the total cost of ownership. Manufacturers aim to develop batteries with higher cycle lives to meet the durability requirements of these applications.

Discharge Power
The discharge power of a lithium-ion cell refers to the maximum amount of power the cell can deliver during discharge. This is an important parameter for applications where high-power output is required. This parameter is influenced by the cell's voltage and the current it can safely provide. Discharge power is calculated by multiplying the voltage of the battery by the current it discharges at that moment:

$$\text{Power (Watts)} = \text{Voltage (Volts)} \times \text{Current (Amperes)}$$
$$\text{Power (Watts)} = \text{Voltage (Volts)} \times \text{Current (Amperes)}$$

There are a number of factors which influence discharge power. A lower internal resistance allows a higher current to flow more easily, thereby increasing the discharge power. High internal resistance can restrict the flow of current, reducing power output. Another variable is SoC. A fully charged battery generally can deliver more power compared to one at a lower SoC. This is usually affected by the slope change in voltage as the battery SoC is reduced. For example, phosphate-based chemistries have a relatively flat curve with minimal change while discharging when contrasted to say NMC cathodes. Furthermore, batteries tend to perform differently at various temperatures. High temperatures might allow for higher discharge rates initially but can accelerate degradation. Conversely, low temperatures can reduce the ability to discharge at high power due to increased internal resistance. Finally, over time as a battery ages, its capacity to deliver high-discharge power diminishes due to the degradation of its internal components.

Charge Profile
The charge profile of a lithium-ion cell describes the specific conditions and characteristics under which the cell is charged from a depleted state back to full capacity. This profile includes important parameters such as the charge rate, voltage limits, and current behavior during the charging process. Understanding and optimizing the charge profile is crucial for maintaining battery health, efficiency, and longevity.

The simplest charge profile is usually constructed of two main phases. Initially, the battery is charged at a constant current. This is typically the fastest part of the charging process. During this phase, the voltage gradually increases until it reaches a predetermined level, often close to the cell's maximum voltage. Once the target voltage is reached, the charger switches to a constant voltage mode. During this phase, the voltage is held steady while the current gradually tapers off as the cell approaches full capacity. This phase is critical to fully top off the battery without overcharging, which could be harmful.

A more advanced charge profile usually employs a multidimensional lookup table to determine the best current dependent on SoC, state of health (SoH), and temperature. From here, a current is calculated and constantly updated as the SoCon battery changes. These calculations are performed by the BMS and then provided to a charger.

The charge profile is crucial for maximizing battery performance, lifespan, and safety. Proper management of the charging process can help mitigate the effects of aging and degradation while ensuring that the battery delivers optimal performance throughout its life. Understanding the charge profile is essential for anyone involved in designing, manufacturing, or using lithium-ion batteries, as it directly affects the efficiency, operational life, and safety of the battery.

Maximum Continuous Discharge
The maximum continuous discharge of a lithium-ion cell refers to the highest current that can be safely drawn from the cell continuously without causing excessive degradation or overheating. This parameter is crucial for determining the cell's suitability for various applications, especially those requiring high-power output over extended periods.

The maximum continuous discharge is typically measured in amperes (A). It can also be expressed as a multiple of the cell's capacity in terms of C-rate, where 1C means the current at which the battery would be discharged completely in one hour. For example, a cell with a capacity of 2 Ah and a maximum continuous discharge rate of 2C can safely provide 4 A continuously until it is discharged in 30 minutes.

There are a few factors that can influence these discharge currents like the discussion above. Lower internal resistance allows for a higher discharge rate by reducing the heat generated inside the cell during discharge. Further, the choice and arrangement of electrode materials significantly impact how quickly electrons and ions can move through the cell, thus affecting the discharge capabilities. Another aspect is efficient heat dissipation. This is crucial when discharging at high rates to prevent thermal runaway and maintain cell integrity and lifespan. Finally, different lithium-ion chemistries have varying capabilities and limits regarding how much current they can safely deliver.

Maximum Pulse Discharge
The maximum pulse discharge of a lithium-ion cell refers to the highest current that the cell can deliver in short bursts or pulses, without causing harmful effects like excessive heat buildup, voltage drop, or premature degradation. This measure is particularly important in applications that require high power for brief periods, such as hybrid or electric vehicles where quick energy bursts are essential.

Unlike maximum continuous discharge, which is sustained, maximum pulse discharge is temporary, usually lasting from a few seconds up to a minute. The duration and frequency of these pulses are crucial factors in determining the safe pulse discharge capabilities. As such the C-rate is much higher than for continuous discharges.

The factors influencing maximum pulse discharge are also similar to continuous discharge. Good thermal management and lower internal resistance allow the battery to handle higher pulse discharges by minimizing the heat generated during these short intense bursts. Furthermore, the electrochemical properties of the battery materials and the physical design influence how quickly the cell can handle ion flow during high-rate discharges. The battery's pulse capabilities to deliver high pulse currents can vary depending on the SoC. The higher, the better especially if the currents are significant. Finally, the temperature will have capability effects. Higher temperatures might temporarily enable higher pulse discharges but at the risk of accelerated degradation. Lastly, exceeding the recommended pulse discharge rate can lead to voltage sag (a significant drop in voltage during the pulse), excessive heat generation, and rapid deterioration of the battery's health. This must hence be effectively controlled by the BMS.

For example, in hybrid electric vehicles, the battery might need to provide a high current pulse to assist with rapid acceleration or to capture energy efficiently during regenerative braking. A sophisticated BMS is crucial in applications utilizing pulse discharge. It monitors the battery's condition during these pulses, ensuring that the operations stay within safe limits to prevent damage and maintain battery integrity over time.

Regenerative Pulse Charge Current
The regenerative pulse charge current of a lithium-ion cell refers to the current used to recharge the battery during brief pulses, typically applied in a scenario where energy is recaptured and reused, such as in regenerative braking systems. This method contrasts with standard charging processes by utilizing short, controlled bursts of current that are derived from kinetic energy converted back into electrical energy during braking or similar processes. During a regenerative process such as braking, kinetic energy is converted into electrical energy via the vehicle's motor, which acts as a generator. The generated electricity is then sent back to the battery in

pulses, which can be significantly higher in current than typical charging currents, depending on the dynamics of the vehicle and the braking event.

Regenerative pulse charging involves capturing energy that would otherwise be lost (e.g. as heat during braking) and using it to recharge the battery. This charging method is intermittent and depends on the occurrence of the energy recovery event. As these events may occur at any time, a higher SoC might limit the amount of current the battery can safely accept, requiring systems to potentially divert excess energy or modulate the charge current. Further influences are the overall health and temperature of the battery will affect how well it can accept pulse charges. Overheated or degraded batteries may not handle high currents safely. Crucially, the BMS ensures that pulse charges do not exceed the battery's capacity to absorb the energy safely. It regulates the charge current to prevent overcharging and overheating.

Regenerative charging enhances the overall energy efficiency of systems by recovering energy that would typically be wasted. Incorporating the feature effectively requires advanced control systems which can increase the complexity and cost of the BMS. A clear drawback however is that frequent high-intensity pulse charging can potentially accelerate battery wear if not properly managed.

Temperature Rise of Continuous Discharge

The temperature rise of continuous discharge in a lithium-ion cell refers to the increase in the cell's temperature that occurs as a result of internal resistance and electrochemical reactions during sustained discharge. This temperature increase is a critical factor in battery management and safety, as excessive heat can lead to reduced performance, battery degradation, and safety risks such as thermal runaway.

The temperature rise of a cell as mentioned is directly related to the internal resistance. As was discussed previously, this tends to increase with age and SoC. Moreover, internal electrochemical reactions that are related to ion flow may be exothermic. Clearly, all of these are directly affected by the degree of discharge. The cell design, especially thickness and outer surface area, will have significant effects on how the heat is removed. Thick cells with limited surface area, for example, high-diameter cylindrical cells will have greater challenges compared to thin long prismatic blade cells.

Typically, temperature sensors placed on cells within modules or packs provide information to the BMS. This can then either provide information to the vehicle to potentially increase coolant flow (air or liquid depending on architecture) or start to provide warnings to the vehicle to reduce current. This would then require the vehicle to reduce performance if possible. Excessive temperature rise is a major safety concern as it can lead to thermal runaway, where an increase in temperature causes a further increase in temperature, potentially leading to fires or explosions. As such, if the certain temperature value is exceeded, the vehicle must come to a complete stop.

Battery Pack

For electric vehicles, sufficient energy and voltage must be available to provide the performance specifications for the vehicle. This is done by housing the cells in a structure called a battery pack (Figure 3.4).

The cells are arranged in series configurations to develop the proper voltage and if necessary, in parallel for sufficient energy content. The main components of a pack are the exterior housing, a BMS, a cooling system, a wiring harness, and cells. The cells may be configured inside modules, which in turn are assembled into series/parallel configurations inside the pack. More advanced systems use CTP concepts

FIGURE 3.4 This photograph shows an electric vehicle LIB with its cell pack exposed. Note the number of cells required to deliver the energy needed to drive the vehicle.
Source: xiaoliangge/Adobe Stock Photos.

which forgo the extra material of the modules providing cost savings and enhancing overall gravimetric energy density. However, it is more challenging from a fire propagation mitigation perspective and offers limited service capability. This section will provide an overview of the BMS and then explore packs constructed of modules or CTP methodologies.

Battery Management System

The BMS is one of the most crucial elements of any battery pack. The main function is to provide safe performance of cells while the pack is in operation.

The BMS will monitor significant system variables such as voltage and temperature from the cells through typically a set of module controllers or slave devices. This is then provided to the BMS master controller. The cell type from a particular manufacturer will have its own set of cell performances including current and voltage, charging limits, charging conditions, discharge characteristics, maximum continuous and pulse power operation, temperature maps with SoC, temperature evolution during charge/discharge, resistance characteristics cell life management and characteristics to long and short-term storage. Each of these must be considered in detail for a thorough design with close attention to safety limits. The BMS will monitor these variables and if excursions occur open a set of contactors essentially electrically disconnecting the battery from the vehicle.

Aside from the safety aspect, the variables can be used to calculate certain state variables highly useful for the vehicle to determine the status of the pack. Communication to the vehicle is accomplished via the control area network (CAN) bus modern BMS are designed to monitor and manage these substates continuously. They use sophisticated algorithms to assess and optimize the battery's performance based on real-time data from sensors measuring voltage, current, temperature, and other critical parameters. The overall state of a lithium-ion battery is an aggregate measure derived from evaluating SoH, SoC, SoP, and SoB. Each of these substates contributes to a comprehensive understanding of the battery's ability to perform its required functions reliably and safely.

PRACTICAL INSIGHTS | CAN Bus and SAEJ1939

A key component of connected vehicles – including electric vehicles – is the communication between various vehicle components and their associated software. This allows for important aspects of the vehicle's systems – including the battery – to be monitored while they are functioning, also known as on-board diagnostics (OBD). SAE's family of standards, referred to by their alphanumeric name, SAEJ1939, constitute an important set of recommended practices that govern this communication, to drive both safety and compliance with various state and federal regulations, including those from the California Air Resources Board (CARB) and the Environmental Protection Agency (EPA). While the standards initially focused on commercial and heavy-duty vehicles, they have subsequently been adopted for other forms of transportation. They provide a model for vehicle communications used across the industry.

PRACTICAL INSIGHTS | Battery Electric Vehicles (BEV) and Data Privacy

With the integration of communications within BEVs and connected vehicles of all types come concerns over the protection of data in those communications. Today design engineers must not only consider the impact of data on vehicle performance but also on the driver – and society's – safety and privacy.

Automakers must adhere to existing regulations based on the region, country, or continent where they sell their vehicles, such as the General Data Protection Regulation (GDPR) and Data Act, as well as the California Consumer Privacy Act (CCPA) in the United States. GDPR primarily focuses on individual data privacy, mandating purpose limitations, data access rights, security measures, and more. It does not address issues related to mass surveillance or data collection by governments or large corporations for security or other purposes. Similarly, China's Personal Information Protection Law (PIPL) imposes restrictions on data transfers. Transparency regarding enforcement and accessibility of trial information post-privacy breaches is deemed necessary.

Regulations are evolving to address complexities in data handling, with emphasis placed on user consent, transparency, and secure design. While GDPR provides a baseline, further specific regulations addressing automotive privacy are needed, including those that limit data retention periods, regulate third-party data sharing, and ensure algorithmic transparency.

Collaboration among industry stakeholders, policymakers, and cybersecurity experts is crucial for creating a consistent global framework. Harmonization efforts can address emerging privacy concerns in the dynamic landscape of BEVs.

In developing countries, such as Malaysia and the Philippines, there are additional challenges beyond simply having regulations in place. Data misuse is widespread, especially through popular apps (e.g. messaging apps and food delivery services). There have been cases of sexual harassment where delivery riders use customer information from these apps without permission, violating their privacy. Alongside regulatory efforts, there is a critical need for education programs to combat these issues. Teaching

> users about the importance of protecting their personal information and promoting responsible data practices is essential. By raising awareness and encouraging privacy-conscious behavior, individuals can better safeguard themselves from privacy breaches and the dangers of data misuse.
>
> Addressing privacy concerns in SD-BEVs requires a multifaceted approach involving regulatory compliance, international collaboration, user empowerment, and education. Regulations such as GDPR, PIPL, CCPA, and others emphasize "privacy-by-design and -default". An example of which was published by the European Commision. Collaboration mechanisms among manufacturers, regulators, and BEV users should be strengthened to ensure effective privacy protection. Continuous efforts to adapt regulations to technological advancements and user needs are essential for fostering trust and ensuring privacy protection in the evolving landscape of SD-BEV technology.
>
> Source: Abdul Hamid, U., "Privacy for Software-defined Battery Electric Vehicles," SAE Research Report EPR2024012, 2024, **https://doi.org/10.4271/EPR2024012**.

State of Charge

The SoC of a lithium-ion cell is a measure of the remaining capacity available in the battery, expressed as a percentage of its total capacity. It is an indicator of how much charge is left in the battery relative to its full capacity when it is completely charged. Essentially, SoC provides a numerical value indicating how much charge is left in the battery relative to its full capacity when completely charged. This is analogous to a fuel gauge for a battery, providing users and systems with vital information on how much energy is available for use.

SoC is typically represented as a percentage, where 0% indicates a completely discharged battery and 100% indicates that the battery is fully charged. Knowing the SoC helps in understanding battery usage patterns, planning recharging cycles, and preventing battery degradation from deep discharges or overcharging.

A number of different methods may determine the SoC of a battery. It may be estimated by measuring the battery's voltage. This usually involves a lookup table that may be adjusted for temperatures. However, this method can be imprecise due to voltage variations under different load conditions. Another method is coulomb counting or current integration. This technique involves tracking the flow of current into and out of the battery over time, offering a more accurate measurement of SoC by accounting for energy consumption and recharge. Some systems use complex algorithms that combine multiple sensors and historical data to more accurately estimate SoC, taking into account factors such as cell aging, temperature, and discharge rates. These methods may also be further combined via data fusion using algorithms such as a Kalman filter (Xie et al. 2023).

Accurately determining the SoC is challenging due to the dynamic nature of battery usage and external factors such as temperature and load. Continuous improvements in sensor technology, data analysis, and battery chemistry are crucial for enhancing the reliability of SoC measurements.

Understanding and managing the SoC is vital for efficient battery utilization, prolonging battery life, and ensuring safety in operation. It allows users and systems to make informed decisions based on the remaining energy capacity, thereby optimizing the performance and utility of lithium-ion batteries across various applications.

State of Health

As batteries age, their performance degrades impacting their capacity, power delivery capability, and impedance increases causing heat loss and efficiency reduction. The SoH of a battery is a measure of this degradation and is crucial for predicting the remaining useful life ensuring safe and efficient operation.

Battery SoH is a measure of the condition of a battery compared to its ideal or new state. It is typically expressed as a percentage, with 100% representing a battery in perfect condition and 0% indicating a battery that can no longer function. SoH encompasses several aspects of battery performance, including:

1. **Capacity Fade:** The reduction in the battery's ability to store charge over time. It is the most common indicator of SoH.
2. **Power Fade:** The decrease in the battery's ability to deliver power.
3. **Internal Resistance:** An increase in internal resistance over time, which can affect the efficiency and performance of the battery.
4. **Self-Discharge Rate:** The rate at which a battery loses charge when not in use.

Several techniques are employed to measure and estimate battery SoH, each with its advantages and limitations:

1. **Coulomb Counting:** This method involves tracking the charge and discharge cycles of the battery. By comparing the total charge input and output over time, the capacity fade can be estimated. However, this method requires precise current measurement and is prone to cumulative errors.
2. **Impedance Spectroscopy:** This technique measures the battery's internal impedance at different frequencies. Changes in impedance can be correlated with SoH. It provides detailed information but requires specialized equipment.
3. **Electrochemical Methods:** These involve analyzing the battery's electrochemical properties, such as voltage, current, and temperature, to estimate SoH. While accurate, these methods can be complex and time-consuming.
4. **Machine Learning Models:** With advancements in data analytics, machine learning models are increasingly used to predict SoH. These models use historical data and real-time battery parameters to provide accurate SoH estimates. They require large datasets and computational power but offer high accuracy and adaptability (Ren and Du 2023).

Advanced data fusion techniques may also be used combining a number of the above-named methodologies. This can provide a number of additional dimensions to the SoH output more realistic to actual behavior. For example, a cell may have high capacity but if the resistance is also high, this will result in reduced performance.

The SoH of a battery is influenced by a variety of factors. These factors affect the battery's performance, longevity, and reliability. The number of cycles a battery undergoes significantly impacts its SOH. Frequent cycling can lead to capacity fade and increased internal resistance. Deeper discharges reduce the battery's lifespan. Keeping the DoD shallow can prolong battery health. High rates can cause overheating and accelerate degradation. High temperatures accelerate chemical reactions within the battery, leading to faster degradation. Low temperatures can increase internal resistance and reduce capacity temporarily. Physical stress from vibration and shock can damage battery internals. Cell chemistry will also affect SoH levels.

The materials used for electrodes (e.g. lithium, nickel, and cobalt) have different degradation mechanisms and rates. Furthermore, this may also be affected by the electrolyte used. Finally, keeping the cells charged at high voltage can also have detrimental effects.

Determining the SoH of batteries in automotive applications presents several unique challenges. These challenges arise due to the demanding operational conditions, diverse usage patterns, and the need for accurate and real-time SoH monitoring. For example, automotive batteries experience highly variable loads due to acceleration, braking, and varying driving conditions. This makes it difficult to apply standard SoH measurement techniques consistently. Batteries in vehicles are exposed to a wide range of temperatures, both environmental and operational (e.g. rapid heating during fast charging or driving). These temperature fluctuations can affect battery performance and complicate SoH assessment.

Automotive batteries consist of multiple cells arranged in series and parallel configurations. The SoH of individual cells can vary (heterogeneous heating, manufacturing differences), making it challenging to determine the overall SoH of the battery pack. Uneven degradation and cell balancing issues can lead to inaccuracies in SoH estimation. Ensuring all cells degrade uniformly is difficult in practice.

The accurate SoH determination method requires precise tracking of all charge and discharge cycles. However, inaccuracies in current measurements and cumulative errors over time can lead to incorrect SoH estimation. Further complicating the matter, batteries degrade due to multiple mechanisms, such as electrode material degradation, electrolyte decomposition, and formation of SEI layers. Modeling these complex and interacting degradation mechanisms accurately is challenging. Developing accurate models requires extensive historical data under various conditions. Collecting and processing this data for diverse vehicle types and usage patterns is resource intensive. Finally, inaccurate SoH determination can lead to unexpected battery failures, posing safety risks. Ensuring high predictive accuracy is crucial to prevent such incidents. Degraded batteries are more susceptible to thermal runaway, which can lead to fires or explosions. SoH monitoring must be sensitive enough to detect early signs of such risks.

State of Function

The state of function (SoF) of a lithium-ion battery is a measure that describes the battery's ability to meet specific performance requirements or functions at any given time. This is unlike SoC, which quantifies the remaining energy as a percentage, or SoH, which indicates the general condition or degradation level of the battery. SoF assesses whether the battery can deliver the required power output under current and anticipated usage conditions. It is a crucial measure for applications where not just any level of charge is sufficient, but where a certain power delivery is necessary for proper functioning. This is especially relevant in scenarios such as needing to accelerate or climb.

For example, a higher SoC generally means more available energy, but SoF specifically concerns whether the power can be delivered as required. As was discussed above, the temperature can have an adverse effect on SoC and as such will also influence SoF. In another case, a degraded battery (low SoH) might struggle to deliver the required power levels, affecting SoF. The calculated SoF will also be directly impacted by the specific demands of the application. As high currents for acceleration in EVs are measured, the SoF must be updated as the energy of the battery is depleted or if the safe temperature of the range is exceeded.

State of Power

The state of power (SoP) of a lithium-ion battery refers to its ability to deliver a specific amount of power at a given moment. It is a crucial metric for understanding the immediate load-handling capabilities of the battery, especially in applications that require high or variable power outputs, such as during acceleration.

SoP quantifies the maximum power output a battery can provide at any point in time, reflecting both the energy available (related to SoC) and the battery's health and functional capacity (related to SoH and SoF). The variable is typically measured in watts (W). Accurate knowledge of SoP is essential for ensuring that battery-powered systems operate within their performance thresholds, preventing power-related failures, and optimizing system responses to varying power demands. Similar to SoF, SoP is dynamically assessed, considering the current SoC, the battery's health, and environmental conditions that might affect performance, such as temperature.

As this measure is a direct function of instantaneous power, a key factor affecting SoP is internal resistance. For example, lower resistance allows for higher power output. Furthermore, the SoC significantly impacts SoP since a lower charge level generally means less power can be drawn from the battery. This means it is immediately coupled to temperature. Performance peaks at optimal temperatures and declines if the battery is too hot or too cold, affecting its power delivery capabilities. Lastly, as the battery ages, its ability to deliver high power can diminish due to increased internal resistance and loss of active material.

Estimating SoP accurately is complex because it must account for dynamic and rapidly changing conditions within the battery and its environment. Advanced sensors and modeling techniques are typically required to predict SoP effectively, integrating data from multiple sources to provide a real-time assessment.

State of Balance

The state of balance (SoB) of a lithium-ion battery refers to a battery pack composed of multiple cells and is a measure of how evenly the charge is distributed across all the cells within the pack. This balance is crucial for ensuring that each cell within the battery pack operates efficiently and safely, extending the overall life and performance of the battery.

SoB reflects the uniformity of the voltage, charge, or capacity across multiple cells in a battery pack. A well-balanced battery pack has cells that all show very similar or identical states of charge. Proper balance is essential because even small differences in cell voltages can lead to underutilization of total capacity and accelerated degradation of certain cells, which can ultimately affect the performance and safety of the entire battery system.

SoB is often determined by measuring the voltage of individual cells within a battery pack since cell voltage is a direct indicator of its SoC. The BMS monitors these voltages continuously and compares them to ensure that all cells remain within close range of each other. Slight differences in the manufacturing process can lead to variations in cell capacity and resistance, which can affect SoB. Furthermore, as cells age or undergo cycles of charging and discharging, their characteristics change differently, impacting their balance. Heterogeneous aging may occur if uneven temperatures are present in the battery pack. This is usually the result of an outside heat source such as the exhaust of a hybrid vehicle or improper thermal management.

There are several methods to correct these problems. For example, passive balancing involves bleeding off charges from cells that are more charged than others through resistors, turning excess electricity into heat. Active balancing moves charge from higher-charged cells to lower-charged ones, often using more complex electronic circuits. This method is more energy-efficient and effective for maintaining balance over longer periods and under more strenuous conditions. With today's advanced manufacturing methodologies, most cell-to-cell differences may be minimized requiring only passive balancing.

State of Energy
The state of energy (SoE) of a lithium-ion battery refers to the total amount of usable energy that the battery can currently deliver, expressed typically in watt-hours (Wh). It provides a measure of the energy content relative to the battery's full energy capacity when it is fully charged. This metric is crucial for assessing how much power the battery can provide for a given duration under specific conditions, and it is particularly important for energy management in systems that require precise knowledge of available energy for effective operation.

SoE is a comprehensive measure that takes into account the SoC and the overall capacity of the battery. While SoC is usually expressed as a percentage, SoE quantifies the actual energy available in absolute terms, such as watt-hours, making it directly relevant for planning and operational decisions.

SoE can be calculated by integrating the voltage and current over time as the battery discharges. This method provides a direct measurement of the energy output and is highly accurate. Furthermore, SoE is often estimated from the SoC and the battery's total capacity. If a battery has a total capacity of 100 Wh and the SoC is 50%, the SoE would be 50 Wh. More sophisticated methods involve using algorithms that account for various factors including battery age, temperature, and discharge rate to more accurately predict SoE.

Design and Structure
Battery packs today may be classified into two different types: modular and CTP. In the next section, both of these methods will be described. In both cases, thermal management will be well considered along with mechanical robustness.

Modular
Battery modules are essential intermediaries between individual cells and the full battery pack. Each module typically contains several cells connected either in series, parallel, or a combination thereof to achieve the desired voltage, capacity, or power requirements. Integrated are sensors both for temperature and voltage. The design may also consist of added thermal propagation mitigation components to help contain any potential safety events. Finally, thermal regulation may be part of the module housing structure or the modules may be thermally bridged to an external cooling plate. The design of these modules is crucial because it directly affects the overall efficiency, safety, and performance of the battery system.

When designing battery modules for automotive applications, engineers must consider several factors. Of crucial importance cascading down the pack level performance to the module. This will determine the voltage, capacity, and mechanical envelope necessary. The choice between cylindrical, prismatic, or pouch cells affects how the cells are arranged within a module. Each type has specific advantages and trade-offs concerning energy density, manufacturing complexity, and thermal management that have been discussed. Care must be taken to consider the swell force the cells will produce. The cells will not only swell as they age but also with changes in SoC. Furthermore, some cell technologies require the cell stack to be placed under mechanical pressure by the placement of a belt. The series and parallel arrangement of cells within a module must balance between achieving higher voltages or increased capacity while ensuring uniform charge and discharge cycles across all cells to maximize efficiency and lifespan. Effective thermal management systems are crucial to maintain optimal operating temperatures. Modules are often designed with cooling channels that allow for liquid cooling solutions, which are more effective than air cooling in managing the heat generated by the cells during operation. This will be highly dependent open the mission profile the vehicle is to achieve. Finally, depending on cell chemistry and packing, the type of fire propagation mitigation strategy must be integrated. This may be as simple as designing specific spacing gaps between cells and adding heat barriers such as mica or combination. Added to safety may also be the addition of fuses in the event of failure by the BMS.

The assembly of the modules is a complex process that involves a number of steps. For volume-level production and to help in quality, these modules tend to be highly automated. Cells are precisely placed and connected using laser welding or ultrasonic welding to ensure strong and reliable connections. These components are critical for distributing current evenly across the cells and must be designed to minimize resistance and maximize conductivity. The cells may be glued or affixed with specialized double-bound tape. To protect against environmental factors and mechanical damage, modules are often encapsulated in durable materials that also aid in thermal management. This final step involves embedding the necessary cooling channels and installing sensors and other BMS components to monitor and regulate the module's operation.

Using modules in the construction of battery packs offers several benefits. Modules allow for easier scaling of battery capacity and voltage. Depending on the vehicle's needs, modules can be added or removed to tailor the battery pack's performance. In case of failure, individual modules can be replaced without the need to dismantle the entire battery pack, reducing maintenance time and costs. Finally, by isolating battery cells into smaller, manageable units, modules help contain potential issues such as thermal runaway within a confined space, preventing it from affecting the entire battery pack.

Despite their advantages, battery modules also pose several challenges. The additional components and assembly steps involved in creating modules can increase the overall cost and complexity of the battery system. Modules add extra weight and take up more space within the battery pack due to the inclusion of additional casing and interconnect, potentially reducing the energy density compared to more integrated solutions like CTP technologies. Finally, maintaining a uniform temperature across all cells within a module can be challenging, especially under high load conditions necessitating advanced cooling solutions.

Cell to Pack (CTP)
As the demand for higher energy densities and more efficient manufacturing processes continues to grow, especially in the EV sector, CTP technology is expected to see wider adoption. The push for better, more compact battery solutions may drive further innovations in CTP, including improved materials and smarter, integrated BMSs.

The concept was driven by the need to enhance energy density, simplify manufacturing processes, and reduce costs. There have been a number of milestones to allow for broader adoption of the technology. The earliest feasibility studies demonstrated fairly quickly the advantages of energy and manufacturing efficiencies. A number of companies stand out in spearheading this technology such as Contemporary Amperex Technologies and BYD. The adoption of regulatory bodies was also important to approve designs based on CTP.

CTP technology involves several technical innovations that distinguish it from traditional battery pack designs. The clear advantage is the elimination of modules, which significantly reduces the number of structural components. The result is a compact lighter design, translating into longer driving ranges for EVs, which is a critical factor for consumer acceptance. This has allowed cells with lower energy density to compete and exceed performance at the pack level with more energetic cell chemistry and format that may require the additional safety of a module. For example, an LFP-based pack using long blade, prismatic cells may have a higher energy density than a pack constructed of high nickel content NMC cells requiring modules. This approach not only improves energy density but also substantially reduces costs. Advanced cooling systems and thermal interface materials are employed to manage heat generated by densely packed cells ensuring safety and performance. Finally, CTP technology optimizes the electrical configuration of the pack, improving energy density and performance. This includes innovations in cell interconnection and power management systems to enhance overall efficiency.

Developing a robust and reliable CTP design requires significant engineering expertise. This includes optimizing the structural, thermal, and electrical aspects of the pack. While CTP can improve thermal management, it also requires advanced solutions to manage the heat generated by densely packed cells. Innovative cooling systems are needed to prevent overheating and ensure safety. Ensuring the safety of CTP battery packs is critical, especially in the event of a crash or thermal runaway. The battery pack cannot rely upon potential safety walls from a module to reduce the possibility of a possible thermal cascade throughout the pack. Lack of industry-wide standards for CTP technology can hinder widespread adoption. Collaborative efforts are needed to establish standards and guidelines for the design and manufacturing of CTP packs. Repairing or replacing individual cells can be more challenging in a CTP design since cells are more integrated into the pack. Additionally, scaling the battery capacity might require complete redesigns rather than simply adding or removing modules.

Several companies have successfully implemented CTP technology in their EVs. For example, BYD's blade battery is a notable example of CTP technology. It offers improved safety, energy density, and thermal management compared to traditional designs. The blade battery is based on elongated prismatic LFP cells. It has been successfully implemented in several BYD EV models, showcasing its commercial viability and performance advantage. CATL's CTP battery packs are used in various commercial EVs. Their technology focuses on maximizing energy density and simplifying the manufacturing process. CATL's CTP packs have been praised for their performance and reliability, setting industry benchmarks. Finally, Tesla has also explored CTP technology for its future battery packs. By integrating cells directly into the vehicle's chassis, Tesla aims to improve energy density and reduce costs. This approach exemplifies how CTP technology can be adapted to different vehicle architectures for optimal performance.

Thermal Propagation

Thermal propagation is a phenomenon where excessive heat generated within a battery spreads to adjacent cells, potentially leading to a cascading thermal event. This can result in catastrophic failures, including fires or explosions, posing significant risks to passengers and property. Understanding the mechanisms of thermal propagation and developing effective mitigation strategies is crucial for the advancement and safety of EV technology.

Effective thermal management in EVs is essential to maintain optimal battery performance and longevity. It involves regulating the temperature of the battery pack to prevent overheating and thermal runaway. Various cooling techniques, including liquid cooling and phase change materials (PCMs), have been explored to manage battery temperatures. Several factors contribute to thermal propagation in EV batteries, including:

- **Overcharging:** Overcharging can lead to excessive heat generation and increased internal pressure, causing thermal runaway and potential thermal propagation.
- **Improper Charging:** High charge rates at low temperatures can lead to dendrite formation with potential shorting in cells.
- **Physical Damage:** Physical damage to battery cells, such as from accidents or impacts, can create internal short circuits, leading to localized heating and subsequent thermal propagation.
- **Manufacturing Defects:** Defects in the manufacturing process, such as impurities or misalignments, can create weak points in the battery cells, making them more susceptible to thermal runaway.

The thermal stability of battery chemistry significantly influences the safety and reliability of EVs. Two commonly used chemistries in the EV industry are Lithium NMC and LFP. These chemistries have distinct thermal characteristics, impacting their performance and safety.

NMC

NMC batteries are popular due to their high energy density and good overall performance. They are widely used in EVs that require long ranges and high-power output. However, NMC batteries have notable thermal stability challenges.

Thermal Stability Characteristics

1. **Operating Temperature Range:** NMC batteries typically operate within a temperature range of −20 to 60°C. However, their optimal performance is maintained between 20 and 40°C.
2. **Thermal Runaway:** NMC batteries are more prone to thermal runaway, a condition where an increase in temperature leads to further temperature increases, potentially resulting in fire or explosion. Thermal runaway in NMC batteries can occur at temperatures above 200°C.
3. **Heat Generation:** NMC batteries generate more heat during charging and discharging compared to LFP batteries, necessitating sophisticated thermal management systems.
4. **Decomposition Temperature:** The decomposition of NMC materials starts around 150 to 200°C, leading to the release of oxygen and further exacerbating thermal runaway.

Lithium Iron Phosphate (LFP)

LFP batteries are known for their excellent thermal stability and safety features. While they have a lower energy density compared to NMC batteries, their robustness in high-temperature environments makes them a preferred choice for applications where safety is paramount.

Thermal Stability Characteristics

1. **Operating Temperature Range:** LFP batteries can operate in a broader temperature range of −20 to 70°C, with optimal performance between 20 and 40°C.
2. **Thermal Runaway:** LFP batteries are significantly less prone to thermal runaway. Thermal runaway in LFP batteries typically starts at temperatures above 270°C, providing a higher safety margin.
3. **Heat Generation:** LFP batteries generate less heat during operation, making them easier to manage thermally and reducing the need for advanced cooling systems.
4. **Decomposition Temperature:** The decomposition of LFP materials begins at higher temperatures, around 300°C, and does not release oxygen, making them inherently safer during thermal events.

However, the typical electrolyte content of LFP cells is higher than NMC for similar capacities. This gives additional fuel that may result in a longer burn time. Moreover, in both cases, temperatures may well exceed over 800°C. Even though LFP may have a higher activation temperature, both chemistries will produce significant heat when in a runaway condition allowing neighboring cells to also ignite.

To mitigate fire propagation in packs there are a number of strategies based on layering safety systems. These start at the cell level, pack, and then BMS. Thermal barriers are materials placed between battery cells to slow down or prevent heat transfer. One common approach is to use aerogels and ceramics. These materials will have low thermal conductivity, which slows the heat transfer between cells. Another material that is used PCMs. These will absorb heat during phase transition, delaying the propagation of thermal runaway. Both of these tend to be lightweight and are relatively easy to implement within a manufacturing environment. However, there is also a cost penalty.

Design improvements focus on cell spacing, module configuration, and the integration of safety features. Increasing the distance between cells reduces the likelihood of heat transfer through conduction. This may be enhanced in conjunction with the above strategy of placing low thermal conductivity material in the space. Further, arranging cells in a way that minimizes heat transfer paths can enhance safety. For example, in systems using cylindrical cells, a honeycomb structure is typically used. Incorporating features such as rupture discs and venting mechanisms can prevent pressure buildup and reduce the risk of explosion. These may be placed on the modules and must be placed on a battery pack. Care must clearly be used where the vent is located to allow free movement of hot gases away from the passenger compartment.

At the pack level, advanced cooling systems include liquid cooling, air cooling, and immersion cooling designed to dissipate heat generated within the battery pack. Liquid cooling provides efficient heat removal and maintains a uniform temperature across the battery pack. A more simpler alternative is air cooling however this approach is significantly less effective due to lower thermal conductivity. A superior methodology is to use immersion. In this case, the battery cells are submerged in a dielectric fluid offering excellent heat transfer.

Finally, a robust BMS monitors the temperature, voltage, and current of each cell, taking preventive actions to avoid conditions leading to thermal runaway. Real-time monitoring allows for immediate response to abnormal temperature rises. For example, the current may be reduced to avoid additional resistive heating in the battery pack. Furthermore, measuring the voltage on every cell provides the necessary feedback to control potential overcharge conditions. In both cases, warning zones may provide initial feedback to the vehicle to either reduce performance (lower speed) or increase the rate of cooling. In worst-case scenarios, the contactors may be closed fully shutting down the battery pack.

Future Considerations

The construction of automotive battery packs is truly a fascinating and challenging multidisciplinary subject area. A strong background in electrochemistry is necessary to select the optimum cell for the mission profile of the vehicle but this is just the beginning. The format of the cell, integration into a module, or going straight CTP is a complex optimization problem requiring strong expertise in mechanical engineering. This is further complicated by fire propagation constraints, thermal development, vehicle integration, crush standards, shock, and vibration. Controls must also be well developed in order for the cells to stay in a very specific operating window for temperature and voltage. The BMS must be able to interpret the data from the cells to determine integrated outputs to the vehicle such as SoC. Most importantly, this is a critical component of the safety topology. Coding standards must be well adhered to avoid software faults. All this must be well simulated to reduce development costs and time. Finally, the system must be built with constant attention paid to stay economically viable and manufacturable.

Great strides in the last two decades have already been made. From the very humble beginnings of Andreas Flocken in 1888, to the EV1 produced by GM in the 1990s we have reached a point where almost every OEM has an electric vehicle in their product portfolio. Amazingly, we are still in the early stages of this transition. As new chemistries are developed with enhanced energy density such as the promise of a solid state, will create another step change in performance and cost. It is rare in one's career to find oneself and such massive transition cycles – something the automotive world has not seen since Henry Ford. As engineers, it is our job to execute this vision safely and economically.

Words to Know

Nominal capacity A standard measure of a battery's energy storage potential.

Wettability A battery electrolyte's ability to wet the electrode, which affects charging capacity and performance.

Cathode During the discharge process, the electrode where reduction occurs (electrons are received) This is usually considered the positive. This reverses in a lithium-ion battery during charging.

Anode During the discharge process, the electrode where oxidation occurs (electrons are lost). This is usually considered the negative. This reverses in a lithium-ion battery during charging.

Graphite Made of carbon and defined as having a stacked layered structure.
Electrolytes This is a compound (typically a liquid) that allows the transfer of ions between the anode and the cathode.
Separator A permeable member to allow the separation of the anode and cathode while allowing ions to pass through.
Energy Density A standard metric that indicates the amount of energy per unit volume (volumetric) or unit weight (gravimetric).
Coulomb Counting A method of measuring current over time to calculate the transfer of energy.

Further Reading

The University of Texas, Austin's Battery Research Group conducts research on both materials and design. **https://batteries.engr.utexas.edu/**

Stanford University's Slac-Stanford Battery Center also researches the impact of new technologies like artificial intelligence on battery development. **https://batterycenter.slac.stanford.edu/**

References

Chen, G., Yan, L., Luo, H., and Guo, S. (2016). Nanoscale engineering of heterostructured anode materials for boosting lithium-ion storage. *Advanced Materials* 28 (35): 7580–7602. **https://doi.org/10.1002/adma.201600164**

Kim, J.H., Kim, S., Han, J.H. et al. (2023). Perspective on carbon nanotubes as conducting agent in lithium-ion batteries: the status and future challenges. *Carbon Letters* 33 (2): 325–333.

Ren, Z. and Du, C. (2023). A review of machine learning state-of-charge and state-of-health estimation algorithms for lithium-ion batteries. *Energy Reports* 9: 2993–3021.

Tong, B. and Li, X. (2024). *Towards Separator Safety of Lithium-ion Batteries: A Review*. Materials Chemistry Frontiers.

Tuo, K., Sun, C., and Liu, S. (2023). Recent progress in and perspectives on emerging halide superionic conductors for all-solid-state batteries. *Electrochemical Energy Reviews* 6 (1): 17.

Xie, J., Wei, X., Bo, X. et al. (2023). State of charge estimation of lithium-ion battery based on extended Kalman filter algorithm. *Frontiers in Energy Research* 11: 1180881.

Yang, T., Wang, C., Zhang, W. et al. (2023). A critical review on composite solid electrolytes for lithium batteries: Design strategies and interface engineering. *Journal of Energy Chemistry* 84: 189–209.

CHAPTER 4

Vehicle and Infrastructure Integration

Oliver Gross
Electrical Energy Technology, Stellantis, Auburn Hills, MI, USA

Introduction

An electric vehicle (EV) has many similarities to a conventional vehicle, with the exception that its infrastructure is completely different. A conventional vehicle that relies on an internal combustion engine (ICE) receives its energy from the liquid or gaseous fuels that are supplied by current infrastructure, most notably local fueling stations. An EV, however, interacts with the electrical infrastructure, receiving electric charge through a charging station. While conventional vehicles fuel via a mechanical operation of transferring a liquid or gas from one storage vessel (the fueling station) to another (the vehicle fuel tank), an EV is "fueled" through the active transfer of electrical energy from an electric network to an electric energy storage device (battery) on the vehicle. The act of fueling an EV is to actively operate the vehicle propulsion system, which is fundamentally different than the passive action of fueling a conventional vehicle.

If the infrastructure actively engages in the operation of a vehicle during recharging, then the charging function of a vehicle cannot be designed or performed without an understanding of the functionality of the infrastructure. Conversely, the needs for the safe operation of the vehicle during charging cannot be placed solely on either the vehicle or the infrastructure. It is therefore a requirement that the interface between the vehicle charging system and the infrastructure electrical supply be managed in such a manner as to protect the integrity of both the vehicle and the infrastructure from its counterpart.

Electric Vehicle Batteries: From Sourcing to Second Life and Recycling, First Edition.
Edited by Bob Galyen and Frank Menchaca.
© 2025 John Wiley & Sons, Inc. Published 2025 by John Wiley & Sons, Inc.

What You Will Learn in This Chapter

In this chapter, we will explore the relationship between the EV and the infrastructure. The reader will learn to:

- Understand the differences between an EV and an ICE vehicle in terms of infrastructure.
- Understand the safety risks that are inherent to EVs and their infrastructure.
- Understand the relationship between the mechanics of a battery EV (BEV) and their impact on the infrastructure considerations.
- Identify the differences between types of EVs and their charging infrastructure.
- Identify and use applicable standards and regulations related to EVs and infrastructure.

Introducing the EVSE and the Battery

We are fortunate that the electrical utility grid is a mature technology, and it represents an expansive infrastructure from which we can support vehicle electrification. For the purposes of discussion, we will focus on the specific infrastructure of interest to be the interface for the vehicle, to the electric utility grid, known in general as EV supply equipment (EVSE). This is the stationary side of the vehicle charging system and interfaces with the vehicle. The charge supply to the EVSE will be high-voltage (HV) alternating current (AC) and will be modified and then transferred to the vehicle, by the EVSE. The electric power supplied will be HV and can be either AC or direct current (DC). Some concerns have been raised over the grid's ability to support large levels of DC fast charging (DCFC) (Elsayed et al. 2024).[1] This is creating a need for improved energy management solutions on the grid side, most notably using artificial intelligence and pairing stationary battery energy storage with the charging side of the infrastructure.

Once charged, the EV battery will need to safely store the electrical energy for extended periods of time and under a wide variety of environmental conditions, in the same manner as done by a conventional vehicle with its fuel. Great efforts have been taken to develop engineering best practices that assure the design, manufacture, and safe operation of EVs. When using vehicle fires as a guideline, EVs are generally prone to fewer fires than conventional vehicles, for total miles driven (NTSB 2020).

An HV electrical energy storage device, however, does have its own challenges that need to be addressed, the most significant of which is the isolation of HV from the battery enclosure and host vehicle, as well as all surroundings. Additionally, isolation failure between elements within the battery system itself is also a challenge to be addressed. Sufficient isolation failure within a battery system can lead to a shorting pathway for the electrical energy of the battery to use, which could provide the conditions for a battery thermal runaway. Battery thermal runaway is a condition where a battery generates excess heat that cannot be removed quickly enough by the battery design, leading to a large release of thermal energy, up to and including fire.

---

[1] https://doi.org/10.1016/j.egyr.2024.05.030

PRACTICAL INSIGHTS | Anatomy of a Li-Ion Battery Thermal Propagation

All batteries can be subject to thermal propagation, but lithium-ion batteries (LIBs) are particularly notable, given their popularity, high-energy density, and use of potentially flammable organic liquid electrolytes. Propagation between cells will occur when there is a source for heat generation within a cell that generates heat more rapidly than the cell can release, leading to exothermic processes within the cell that generate significantly more heat, leading to rapid heating of adjacent cells to the point where they too experience the same undesirable exothermic processes and continue to propagate to their neighboring cells. The by-products of this event not only include heat but also include hot gases, smoke and particulates, and fire.

The initial source of heat can be internal to the cell or external. Most thermal events originate with an electrical short circuit. The short circuit can be a result of a cell or pack design flaw and/or a process defect. In HV packs, isolation failures are a common cause, if the failure creates two separate points of isolation loss, allowing for an external short circuit. The electric current created within the short circuit must be high enough and be sustained long enough to generate sufficient local heat to initiate internal cell exothermic events. Examples of some critical temperatures are as follows:

(A) 85 °C – Electrolyte decomposition of the anode solid electrolyte interphase (SEI) layer, allowing the electrolyte to react with the anode active material.

(B) 115 °C – Decomposition of high-nickel cathode materials, which in turn release heat and oxygen.

(C) 130–150 °C – Melting collapse of separators made from polyethylene or polypropylene, which can include shrinkage that exposes edges of cathode and anode electrodes.

(D) 180–200 °C – Decomposition of iron phosphate cathode active materials, along with oxidation of anode material.

The electrolyte solvents will begin to decompose at lower temperatures, generating gases that are normally vented from the cell. The gases can carry with them a great deal of thermal energy, and a good pack design will both disperse the gases and vent them off board the battery pack. Reaction times within a shorted cell can be rapid, meaning the reaction will progress to higher temperatures, and release other reaction products, many of which are solid particulates.

Hot solid particulate materials will not only be carried with the hot vent gases but can also be deposited on other surfaces within the battery pack. They can in turn cause local isolation failures, which induce a secondary failure pathway within the pack, further expanding the pack propagation. During the dilution of the hot gases within the pack, these gases will mix with the local air within the pack and can be at or above the autoignition temperature for those gases. Good pack design would rapidly dilute the hot gases so that they fall below the autoignition point, but the presence of hot particulates can further encourage autoignition.

A good pack design will eliminate the electrical current from a pack experiencing a thermal event. It will also provide thermal barriers for cell-to-cell thermal propagation, protection of critical HV connections from particulate contamination, and good gas flow management. The cells themselves will utilize methods to couple their heat to the pack thermal system and busbars, to act as heat sinks, along with the use of internal cell technologies to blunt heat generation and propagation.

> **PRACTICAL INSIGHTS | EV Fires in Florida**
>
> Real-world incidents of vehicles exposed to flooding have shown that, despite best efforts, some incidents in HV system isolation do occur, and these can lead to vehicle fires. Recent examples include vehicles that were flooded in Florida under Hurricane Ian (2022) and Hurricane Idalia (2023). Many vehicles that were expected to have sufficient seal protection from water immersion did ultimately fail and undergo thermal runaway, leading to fires (Byington 2022; Impelli 2023; Crowley and Weise 2023).

Not all hazards can be attributed to HV systems. The proliferation of high-energy-density LIB technology has also brought its own challenges to mobility and electrified vehicles. The particularly high volumetric and gravimetric energy density (e.g. Wh/l and Wh/kg) of Li ion means that electric currents during operation can become significant, especially if the cell is abused or an internal or external short circuit is produced. These high currents generate levels of heat energy sufficient that the cell cannot transport this heat to the outside environment quick enough to avoid its internal temperature rising to a point where materials within the cell exothermically decompose, at which point significantly more heat is released, leading to cell thermal runaway.

Lithium-ion cells will release a great deal of heat, smoke, and gases during a thermal runaway event. This is of particular concern when the vehicle is itself located in confined areas or adjacent to other ignitable materials. Even relatively small or modest battery packs can generate large quantities of smoke, heat, fire, and ejected materials, when abused or when containing a latent defect.

The issue of fires associated with e-bike batteries has been especially pronounced in New York City, where the reported number of fire incidents rose from 30 in 2019 to 268 in 2023 (Gelinas 2023; Borja 2024; Dutton 2024). While this is alarming, the fact that e-bikes will be transported on public transit and stored in apartments and other high-density residences makes such fires particularly concerning, leading to a long list of reported severe injuries and deaths.

BEV batteries are considerably larger than those in e-bikes. As a result, BEVs will normally be parked in open areas or parking structures. Many of these locations will permit charging of the vehicle while parked. This is particularly the case for large-scale vehicle fleets and heavy-duty applications, such as buses and trucks. These fleets have recently created new infrastructure demands for extremely high (megawatt-level) charging supplies. Electrification of aviation is in its infancy but is rapidly building upon similar infrastructure demands.

> **PRACTICAL INSIGHTS | E-Flight**
>
> Of all the modes of mobility, aviation is the most difficult to electrify. It is power-dense, specifically requiring high specific energy forms of energy storage. Fortunately, electric motors have high power densities, rivaling gas turbines, although the same cannot be said for batteries or fuel cells. Nonetheless, there has been considerable progress in recent times.

> Aviation is a highly regulated industry. In the United States, it is regulated by the Federal Aviation Administration (FAA), a division of the U.S. Department of Transportation (DOT). There are separate Federal Aviation Regulations (FARs) for aircraft of different sizes. Many have recently been modified to accommodate the advent of electrified aircraft, including electric vertical takeoff and landing (eVTOL) air taxis.
>
> Most flights follow a typical mission profile of takeoff, climb to cruise altitude, cruise, descent, approach, and landing. Some ground maneuvering is also considered. Regulatory requirements will also require the inclusion of an aborted landing, diversion to a second landing site, and a second landing. For aircraft under instrument rules, the diversion time will be at least 45 minutes (Elsayed et al. 2024). For eVTOL air taxis, this profile has been slightly modified, to shorten the reserve for an alternative route. It is this mission profile that will determine the total amount of onboard energy required.
>
> Most batteries will not have constant charge and discharge power over their state of charge (SOC) range, with discharge power declining with declining SOC. This power restriction needs to be accounted for in propulsion system sizing, particularly when accounting for reserve and seeking an alternate landing. It is common in eVTOLs to commit less than 40% of the total SOC range to the standard mission profile. This means that a cell with a rated specific energy of 250 Wh/kg will in practice behave as if it is operating at 100 Wh/kg.
>
> Battery energy density and power density are challenged by these mission realities, and this has led to a technology roadmap for e-aviation that is distinct from other forms of mobility. System integration methods have assisted in making the best use of current cell technologies so that practical short-range eVTOL and commuter aircraft are possible. Propulsion systems have also moved to higher system voltages than most ground vehicles, such as 1,200 V, allowing for the minimization of required electrical current and thermal losses. Total battery energy sizes of 0.5–3 MWh are envisioned for many of these aircraft, necessitating consideration of megawatt charging (MWC)-level recharging.
>
> Battery life and duty cycle for e-aviation are also markedly different from other mobility profiles, given the rapid accumulation of shallow cycles and high relative duty cycles. This will mean that battery service and replacement will also become a significant part of the lifecycle, requiring particular attention to battery recycling.

Bidirectional charging for electrified vehicles is also a rapidly evolving technology, where the vehicle is seen not only as a means of mobility but also as a source of electrical energy. Charging infrastructure has been rapidly developing to ensure efficient and safe power exchange between the vehicle and stationary sources, such as homes, buildings, and the grid. While bidirectional charging presents a great advantage for the widespread use of EVs, it also brings challenges in the areas of control and fault detection.

A vehicle battery's interaction with infrastructure is not just limited to the confines of the vehicle and in its normal use state of parking, charging, and driving but also to long-term storage or transport of the battery, either in the vehicle or separate from the vehicle. Given all the unique characteristics and consideration we have extended to the battery, it is reasonable to expect the same holds true for these cases. Recent experiences on thermal runaway of batteries and vehicles in transport reinforce a need for extended diagnostics, communication, and mitigation for battery incidents. In an effort to circumnavigate the strict transport regulations on lithium and LIBs, some may opt to not appropriately declare their cargo's contents. It has been estimated that around 5% of shipping containers will include undeclared hazardous materials, which greatly complicate the ability of transport agencies and first responders to mitigate hazards and threats associated with these goods (Schuler 2022; Editorial Team 2022; Glavan and Burke 2022).

The advent of large-scale adoption of lithium and lithium-ion traction batteries has begun to rescope our interpretation of infrastructure, with respect to our mobility devices and vehicles. This in turn requires that we think differently about the requirements a battery system will need to adhere to when interacting with its environment and infrastructure. In this chapter, we will explore these requirements and the resultant engineering responses in design.

Electric Vehicle Types

The use of electric propulsion is straightforward, particularly when compared to conventional ICE-powered vehicles. Not all electrified vehicles are purely electric, however, so it is good to first understand the similarities and differences between these propulsion systems. Those electrified vehicles that require some form of recharging from the electrical infrastructure are generalized into the term plug-in EV (PEV).

Battery Electric Vehicle

A BEV is a vehicle that derives all traction from an electric motor, which is supplied purely by an electrical energy source. All electric mobility devices that use a battery to exclusively propel their devices can therefore fall into this category. For any electrified ground vehicle, the propulsion system can be described as shown in Figure 4.1. Here, the battery is the electrical energy storage device that stores the electrical energy, provides and receives electric power to and from the inverter, and provides and receives electric power to and from a charger. The inverter converts the DC from the battery to AC (often 3-phase) that is used by the electric motor to drive the vehicle. During braking, the motor can switch to being a generator, where it will generate AC, and the inverter rectifies this AC to DC, so it can be accepted by the battery.

What is also important to note is that the charger may be on-board the vehicle or off board the vehicle. Most BEVs manufactured by automotive original equipment manufacturers (OEMs) will include an on-board charger (OBC) that will accept electric current from external supply equipment, rectify that AC to DC, and, at the

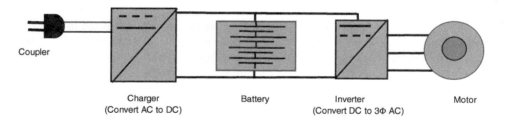

FIGURE 4.1 Schematic diagram for a battery electric vehicle powertrain.

correct voltage for the battery, supply that power to the battery. Smaller portable BEVs, such as e-bikes, scooters, and hoverboards, may use an external charger to perform these functions. Similarly for larger BEVs using FC, this rectification and supply will be off board the vehicle.

Hybrid Electric Vehicle

A hybrid EV (HEV) is a vehicle where traction is supplied by more than one power source. Most commonly, this is a combination of an ICE powertrain and an electric powertrain. In the most common examples, an HEV does not rely on any external electrical infrastructure and all electric energy is supplied on-board the vehicle. There are multiple ways to blend the ICE and electric power sources, however, and the simplest categorization method is to identify whether the electric power source is in parallel with or in series with the ICE power source.

An example of a parallel HEV is shown in Figure 4.2. The electric propulsion system appears like the BEV, except for the omission of the charger. The ICE powertrain contains a fuel tank that supplies fuel to the ICE, which drives a transmission. That transmission shares an input from the electric drivetrain, where the electric motor will also provide a mechanical torque. The torque output from these two sources can be blended and delivered for traction.

Given this simplified configuration, the engine can deliver all traction torque, the electric motor can deliver all traction torque, or both can share in providing traction torque. Looking more closely, one can also see that the engine can produce additional power that can be supplied to the motor, acting as a generator, and supply electric power back to the battery. Regenerative braking, where the motor acts as a generator to slow the vehicle, can also be performed, where all braking torque is directed to the electric motor/generator. In this manner, all electric energy supplied by the battery

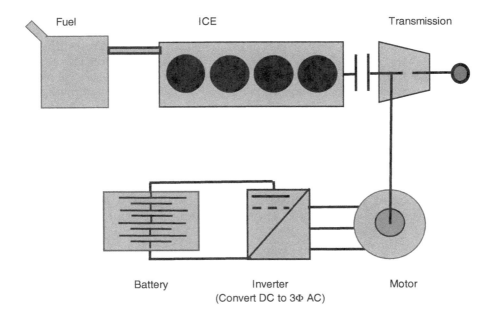

FIGURE 4.2 Schematic diagram for a parallel hybrid electric vehicle powertrain.

can be replenished through a combination of regenerative braking and excess engine power. This way, no external electrical energy supply is required to charge the battery.

> **PRACTICAL INSIGHTS** | Regenerative Charging
>
> One significant benefit EVs have over ICE-powered vehicles is the ability to use regenerative braking to recapture vehicle kinetic energy, rather than dispersing the energy through friction braking. Regenerative braking can be a significant contributor to extending vehicle range and often favors vehicles with higher weights, particularly in urban conditions. There are some limitations to regenerative braking, which can lead to additional losses. When a battery is at a high SOC, or at an operating point where it has a relatively high relative internal resistance, it will be limited to the amount of power and the amount of energy it can accept during the regenerative braking event. Under these functional conditions, the BEV must use foundation/friction braking in support of regenerative braking. The battery management systems will also be challenged under these conditions to accurately manage the real-time battery energy and SOC, placing the battery at risk for degradation or damage. Another control-limited condition may occur at very slow speeds, below 5 mph, where the EV motor/generator cannot reliably manage the vehicle torque from this speed to a stop. There continue to be significant improvements in battery and propulsion system hardware and control, to expand the operating window for regenerative braking use.

Propulsion system functionality in a parallel HEV uses the battery and the electric motor to assist the ICE in propulsion during high-power accelerations, so that the engine can operate in its most efficient power range. In some HEVs, the battery and electric propulsion system are large enough to solely propel the vehicle at low and modest speeds. Additionally, during stops, coast downs, and other low-energy maneuvers, it is possible to switch off the ICE and have the electric powertrain propel the vehicle. There will be acceleration and drive maneuvers that give the engine the opportunity to some excessive power to generate electricity through the motor/generator and recharge the battery, along with the regenerative braking opportunities. It is with these maneuvers that the fuel consumption of the ICE can be curtailed through the use of the electric powertrain.

Variations of this parallel hybrid architecture are very popular among light-duty vehicles and make up the vast majority of HEVs available on the automotive market.

A series hybrid is a different arrangement, which is shown in Figure 4.3. In this configuration, the ICE exclusively drives an electric generator, producing electric power. That electric power is rectified into DC that is supplied to an electric power bus, which is connected to both a battery and an inverter. The inverter is connected to

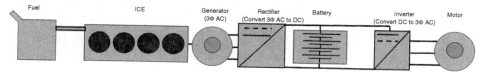

FIGURE 4.3 Schematic diagram for a series hybrid electric vehicle powertrain.

the electric motor/generator, and all traction is supplied by the electric motor operation. Functionally, this arrangement is very similar to the parallel hybrid except that there is not a direct mechanical connection of the ICE to the driving wheels. The ICE can be configured and operated in such a manner that, when it operates at its most efficient operating point, all transient vehicle maneuvers can be managed by the electric powertrain. This powertrain will normally have a larger motor/generator and inverter than in the parallel case, and the battery is typically larger as well, to account for the larger transient power and energy demands.

Plug-in Hybrid Electric Vehicle

The plug-in HEV (PHEV) is a derivative of the HEV architecture, where the battery energy source can come from both the ICE (as in the HEV) and externally (as in the BEV). The PHEV can expand the all-electric operation over the HEV, by utilizing a battery with usable battery energy (UBE) considerably larger than in the HEV. This energy can be depleted during operation until a critical SOC is reached, below which the battery would not have sufficient discharge power to support all HEV functions. A PHEV adds an OBC to the powertrain, converting AC electric power to DC power, for recharging the battery.

Range-Extended Electric Vehicle

A range-extended EV (REEV) is a type of PHEV but can be a similar expansion of a series hybrid propulsion system. Here, an ICE acts exclusively as an electric generator and is combined with a BEV architecture, where the battery is the primary supply of energy. The ICE extends the vehicle range by generating electric power and supplying it to both the battery and the motor, when the battery SOC is depleted.

Summary

The major differences between these popular propulsion system architectures are summarized in Table 4.1:

TABLE 4.1 Major Characteristics of Comparable Electrified Powertrains

	Battery (Size, Power vs. Energy)		Fuel	ICE	Rectifier	Inverter	Transmission	Motor	Charger
Parallel HEV	Small	Power	X	X		X	X	X	
Series HEV	Small	Power	X	X	X	X		X	
PHEV	Medium	Balanced	X	X		X	X	X	X
REEV	Medium	Balanced	X	X	X	X		X	X
BEV	Large	Energy				X		X	X

How Infrastructure Drives Battery and Vehicle Design

Infrastructure can be considered the material entities from the surrounding ecosystem that directly interface with our systems. This may be seen in the adjoining systems that our system of interest mates with. For a battery system, this would be the vehicle, and for the vehicle, this would be the roads, parking locations, storage, and fueling infrastructures.

For the designer of an electrified vehicle, the major physical aspects of interest related to infrastructure are the distance between locations and the available charging power. In turn, vehicle range will play a significant role in determining the ability to comply with the designer's selected allowable distance between charging events. The time for a recharge will also be dependent on the available charging power, relative to the battery size on-board the vehicle.

The regulatory range for a PEV has been defined by the United Nations Economic Commission for Europe (UNECE) in the United Nations Global Technical Regulation No. 22 (UN GTR No. 22), named In-vehicle Battery Durability for Electrified Vehicles. The term used for the regulatory range is certified range (CR), and the fraction of range available to the vehicle over its life is a state of certified range (SOCR). The Regulation proceeds to define the method used for the determination of certified energy (CE), and how this is tied to the certification drive cycle suite that is selected by each national regulatory body, as well as identifying the default certification drive cycle as the Worldwide Harmonized Light Vehicles Test Procedure (WLTP), defined under UN GTR No. 15.

The vehicle designer will use the certification drive cycle to determine the required UBE to achieve the target range. The UBE is then used to determine the range indication for the BEV, both as SOCR and to support the algorithms used to determine real-time range. The UBE reflects the vehicle demand energy (VDE) that is determined by the physics of the vehicle. While UBE is usually denoted in kilowatt-hour (kWh), the VDE is often normalized to distance traveled under the certification drive cycle, Wh/mi, or Wh/km, which simplifies the battery sizing exercise. VDE plays a significant role in battery sizing, as a high VDE requires a larger UBE for a given range than a low VDE.

The VDE is an integration of the road load power and the kinetic power (Dekraker et al. 2017). Road load power is required to overcome the resistance of the vehicle and can be described for a vehicle through the use of three road load coefficients (F, in N): A (kg), B (kg/(m/s)), and C (kg/(m/s)^2). These coefficients will normally be determined during a vehicle coast-down test. Kinetic power is the power required to accelerate the vehicle and is directly associated with the mass of the vehicle. For a given distance traveled, the simplest VDE equation can be written as follows:

$$\text{Vehicle Demand Energy} = \sum_{i=1}^{N}\left[\left(\text{mass} + A + Bv_i + Cv_i^2\right) \cdot \text{Distance}_i\right]$$

It can be seen that VDE is strongly dependent on vehicle mass and vehicle speed (v_i). The resistive forces that affect the coefficients include friction between

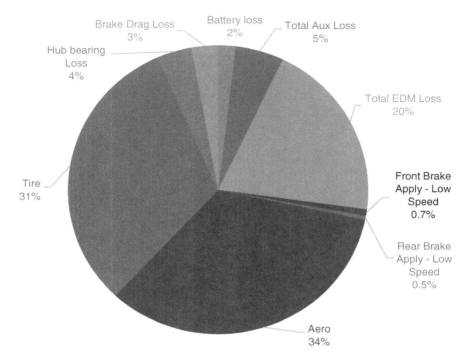

FIGURE 4.4 Typical sources for energy loss in a BEV.

the vehicle and the road, internal friction between the driveline and the wheels, and aerodynamic resistance.

Vehicle mass is the single largest determinant for VDE although aerodynamic losses are not insignificant. Road load losses in a BEV will account for approximately 70–80% of all losses, while the remaining losses originate from the propulsion system and auxiliary system losses, as shown in Figure 4.4.

Managing Weight and Its Effects

Mass remains a significant concern for BEVs, particularly given that the battery mass contributes to a measurable increase in total vehicle mass, negatively impacting VDE. An illustrative example is provided in Table 4.2. All have been normalized to a 300-mile CR. The light car experienced approximately 11% mass increase in curb weight when moving to BEV, while the light truck saw a larger overall mass increase, although a smaller percentage in curb weight increase.

Mitigation strategies for the net VDE increase will approach the reduction of total vehicle mass, the improvement in propulsion and accessory system efficiency/load demand, or a combination of both.

Considerable attention is being paid within the battery engineering community to improving battery-specific energy (Wh/kg), which will in turn lighten batteries. Approaches are diverse, including addressing new battery electrochemistry, novel cell and pack design, and improved management of embedded battery energy. Battery-embedded energy is a measure of the total rated energy that is installed within a battery system. This is normally greater than the UBE, since it needs to account for

TABLE 4.2 Comparison of Mass Impact of BEV Propulsion System on Motor Vehicles

		Light Car		Full-Size P/U Truck	
		ICE	BEV	ICE	BEV
Curb Weight	lbs	3,100	3,834	6,000	7,209
	kg	1,407	1,743	2,724	3,277
Fuel Economy (Gasoline/EV)	mpg/mpge	53	125	19	70
VDE	MJ/mi	0.65	0.97	1.3	1.7
Certified Range	mi	300			
Gasoline Mass Required	kg	15.5		43.3	
Equivalent Battery (160 Wh/kg)	kg		506		903
Mass Savings on BEV Driveline	kg		−170		−350
Net Mass Difference	kg		336		553
	% vehicle		11%		9%

design and control factors such as cell-to-cell capacity variation, system voltage, and current measurement inaccuracies.

Advanced materials and structures offer the promise of both vehicle and battery lightweighting. There are weight savings to be achieved when vehicle and battery structures can be shared or merged, avoiding structural and functional redundancies. Examples of such architectural efforts include cell to pack, where redundant modular components are eliminated, and pack to body, where the battery pack itself becomes part of the vehicle, such as the cabin floor. Careful use of multi-material approaches, where the use of multiple steels, aluminum, and composites is combined, has shown promise in total weight reduction, although factors such as cost, service, and recyclability need to be considered.

Other Infrastructure Considerations for EVs and Their Batteries

While the additional weight a battery brings to the EV can impact VDE, it also has an impact on vehicle dynamics under crash conditions. Additional vehicle mass means additional kinetic energy that must be managed during an impact. The battery itself needs to be designed to protect the internal battery cells and modules from hazards that could result in thermal runaway and to mitigate the effects of thermal runaway if it occurs. While this is a source for some of the battery's weight, the vehicle designer might choose to share these safety functions with the rest of the vehicle structure. The result is additional mass needed in the vehicle, beyond that incorporated in the battery.

Engineering best practices have been established by developers to account for the series of hazards that are unique to EVs. Such hazards include HV, vent products and smoke, heat/fire, and stranded energy.

HV is an inherent hazard that is managed at the vehicle and sub-system levels within an EV. For the HV battery in an EV, this is identified as the threshold where the DC voltage for the battery reaches 60 VDC. At and above this threshold, best practices have the cell string isolated from the battery enclosure, and this isolation is monitored via an integrated measurement circuit. The functionality of the circuit is provided by international and domestic regulations, such as UN ECE R100, and generally evaluates the isolation between the battery string positive and negative terminals and the battery enclosure. In a crash event, it is possible to have the isolation failure of one terminal to the enclosure, and the HV hazard can be contained. A two-point isolation failure, where a return path for current is created, can be potentially created if not accounted for in the vehicle integration. Prevention of the creation of a two-point isolation failure within the battery during crush is key to design the best practices for the battery pack, to ensure a low likelihood of such an event under most crash conditions.

Both vent products and smoke are possible if there is an internal isolation failure or a cell is deformed to a point where it fails. Battery designs will manage the routing of vent products to specific venting points within the battery pack. Those locations are selected in concert with the vehicle design, to ensure the furthest distance from the vehicle occupants or from other potential hazard sources, such as external HV lines or other hot propulsion system components. Vent products and smoke can have toxicity and health hazards of their own, so it is critical that the developer understand the composition of their expected products. Most products are of modest concern, but LIBs can produce levels of hydrofluoric acid (HF) that can potentially be of concern to those that come in direct contact with a large-scale battery failure event.

Heat and fire are always risks from a vehicle crash. The LIBs in EVs can generate heat if an internal short circuit occurs within the battery. The internal short is an isolation failure and can occur at the cell level or higher. If the heat is locally significant, the cell can undergo thermal runaway and potentially lead to fire. Because of the materials used in an EV battery, there will be a significant amount of energy that can be released during thermal propagation. These fires can be intense and sustained, although first responders are well trained to handle such fires.

Stranded energy is a hazard unique to batteries and capacitors. An electrical energy storage device is an active energy storage device, meaning it needs to operate in its primary functional mode (electrically discharge) for its energy to be removed. In a crash event, it is common to disconnect the battery from the propulsion system as part of the de-energizing process. This means that the energy within the battery remains in place after the crash event. Safely and reliably removing this energy can become a challenge if the balance of the vehicle or the battery itself has been significantly damaged by the crash. Most batteries produced by major OEMs can be supported with special means to de-energize cells and modules within the pack, after a crash event, although the battery should be considered hazardous until it has been de-energized.

Given the concerns related to these hazards and the significance that a thermal event can bring, the regulatory approach has been to develop effective monitoring and notification methods, so that the occupants have ample time to get the vehicle to a safe location and get clear of the vehicle. UN GTR No. 20 has set the time from the initial of a thermal event to where the products of that thermal event (vent products, smoke, heat, or fire) can reach the occupants to be a minimum of five minutes.

Design for battery clearance and deformation is a central design activity when integrating the battery into the vehicle. In addition to meeting regulatory crash test minima, adequate clearance for battery pack deformation will be accounted for in the design. As many EV batteries are mounted under the vehicle, the collision or impact of debris or other objects with the underside of the vehicle must be taken into account, to ensure the adequate functioning of the battery through most events. This is normally achieved through both static and dynamic load tests, as well as impact tests, and designing adequate deformation clearance for a given load or impact energy.

There are a range of considerations for siting a PEV charging station, as well as best practices that should be accounted for in the PEV. A charging station should be in an area that is accessible to first responders, who would require access to a vehicle that is involved in a charging incident. Critical not only to first responders but also to the public is the vehicle's own diagnostic and notification systems, which need to announce both physically and electronically the presence of a hazard. This is done not only through means of personal electronic applications but also through the activation of lights and audible alarms on the vehicle. Automotive OEMs also collaborate with first responders in providing easily accessible response information that allows responders to utilize the safety features specific to the vehicle, to counter the incident. Not all of these can be met in all circumstances, particularly when considering single-family dwellings.

Ride and Handling

The additional weight of the battery and its frequent low location within the vehicle often offer benefits to the handling characteristics of a BEV, although it does challenge acceleration and braking, leading to the issues described previously. Dedicated EV architectures have taken advantage of these characteristics, which has led to delightful driver experiences.

Electrification of vehicles has also meant a substantial increase in vehicle accessory loads, including all systems in support of connected and autonomous vehicle (CAV) function and in-vehicle infotainment (IVI). These higher loads do not only reflect in the VDE but also in stationary (e.g. "off-cycle" and "key-off") vehicle operation and use. As shown in Figure 4.1, these loads are treated as losses within the overall system and, as such, need to be computed within the overall UBE determination strategy for the vehicle. Intelligent management of all users on-board the electrical and electronic (EE) powernet of the vehicle seeks to minimize losses by maximizing the efficiency of the energy used.

Aerodynamics

A reduction in overall vehicle drag improves vehicle efficiency and VDE. Drag is reduced by lowering both the frontal and areal drag coefficients for a vehicle, as well as by reducing its frontal area. One unique consideration for BEVs is that the location of the battery under the vehicle has led to generally taller vehicles, meaning a higher vertical displacement of airflow and a larger frontal area. In EVs, aero losses are the single most significant loss, since EVs have a much reduced level of loss from their driveline. For this reason, aerodynamic improvements are of particular benefit to EVs.

Thermal Management

Thermal systems on EVs are different than those on ICE vehicles but are in some ways more critical. The battery on an EV prefers to operate within a modest temperature range, typically 15–45 °C, and thermal systems will be configured to manage the battery within this range, to ensure maximum probability that the UBE can be supplied. A battery will see excursions from this range, and thermal systems will be designed to optimally return the battery to its preferred temperature range, while balancing heating/cooling power, time, and temperature variation within the battery.

Thermal management systems have seen the introduction of more efficient technologies such as heat pumps, thermal energy storage materials, phase change materials, and heat pipes, with the goal of stabilizing the battery temperature while using the minimum amount of battery or infrastructure energy to do so.

Assessing Range Requirements for Current and Future Infrastructure

While the UBE depends on the VDE, both are ultimately impacted by the selection of a vehicle's appropriate CR. A minimum EV range remains non-regulated, apart from the definition of an EV able to achieve zero-emission vehicle (ZEV) credits under California Air Resources Board (CARB) Advanced Clean Cars II (ACC II) provisions, which defined the minimum range to be 200 miles under their certification drive cycle. While daily trips are relatively modest (16–37 miles per day), a vehicle is purchased to also perform longer trips as needed. The charging interface initiative e.V. (CharIN) had originally proposed all interstate highways provide a minimum charging coverage of FC every 125 miles, then reducing that distance to 50 miles.

Current infrastructure plans focus on stationary physical installations. In this manner, the optimum vehicle range (and hence battery size) will depend on infrastructure availability and capability. Availability is dependent on the number of charging locations and their reliability, and the capability of the charger hardware's performance and reliability.

Future infrastructure may move to dynamic wireless power transfer (DWPT), where a vehicle can be charged while moving on a road. In this case, the maximum required vehicle range could be appreciably reduced, and the resultant battery could be downsized. Such infrastructure is currently in its infancy.

1. New Infrastructure Considerations for EVs

The EV is designed with consideration for the infrastructure within which it will operate. The charging infrastructure is of particular interest for the engineer working on EVs and their related batteries, due to its criticality in enabling acceptable use of the EV. To this end, it is important that the EV be simultaneously matched with its use case, or mission, and the limitations of the technologies to be employed. There are many kinds of EVs, and this diversity will continue to expand as new use cases come into being. The electrical infrastructure is more rigid, as it must ensure a reliable and predictable supply of service. Crossing both considerations is the battery technology itself, which not only has a large degree of adaptability and performance but also has its own limitations, which must be accounted for (Tefft 2022).

The battery energy storage system is separate from most of the infrastructure hardware that the EV will face. The following interface diagram (Figure 4.5) shows the layers between the energy storage source (the cells) and the infrastructure itself.

The infrastructure will supply electrical power in some form of AC. There will be some form of interconnection between the vehicle and its related equipment, which is made at point A. Some form of AC-to-DC conversion will occur, along with any required DC voltage and current modulation necessary to support the energy storage system. There are two possible interface conditions: The first is the example shown at junction B, where the conversion takes place off board the vehicle, and

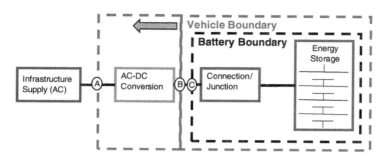

FIGURE 4.5 Infrastructure–energy storage layered interface concept.

therefore, the equipment between A and B is dedicated service equipment for the vehicle, and the second is the example shown at junction C, where the conversion output meets the battery pack. The battery itself contains the energy storage elements of cells or modules, along with a connection/junction that intelligently connects the string to the balance of the vehicle and the system.

The schematic diagram above exemplifies the most common hardware components that create the pathway between the energy storage devices and the infrastructure supply. These can be divided up in multiple different manners, depending on the demands of the use case and the capabilities of the underlying hardware.

System Safety

In almost all cases, the infrastructure supply will be HV, assuming this is defined as >30 VAC. The infrastructure supply can be a simple AC wall outlet or a much more sophisticated connection to grid-tied resources. The infrastructure supply side is stationary and is subject to the regulatory requirements for any such installation, including local building codes. The connection at A is the first point where supply energy will be transferred to the vehicle system. The transfer can be made conductively or inductively, and that interface must suit the voltage range, current range, and environmental conditions expected for that interface. Information may also be passed in concert with that interface.

Since the energy storage system is a DC device, power conversion from AC to DC must be provided. The output DC voltage and current must match the requirements of the energy storage system; otherwise, an additional DC conversion will be required. If the DC output is also HV (i.e. $\geq$60 VDC), isolation of the active circuit from the balance of the system must be required, and correct isolation measurement methods need to be applied.

Voltage control in such systems is not only managed through the AC–DC conversion unit but also monitored by the battery system. Feedback between the battery system and the conversion unit is essential to ensure safe operating voltages for the battery. The connection/junction device enables the battery pack to disconnect if the supply voltage is outside of the acceptable range for the battery. This junction may use simple electronic switches or electromechanical relays. More sophisticated breaker-type switches may also be considered.

Current control for charging carries unique challenges. All DC components within the circuit will have a finite level of resistance, as will every interconnect. The thermal limits of each component and interface need to be known to ensure safe operation. Failure to maintain current control can lead to component overheating, leading to component failure, potential system failure, and loss of HV isolation. The battery connection/junction would be designed to ensure a correct response in the event of overcurrent. In addition to the switches, fusing can be employed. This can be with thermal fuses or more intelligent or precise devices. It is common to select switches or relays that cannot respond to extremely high currents so that the thermal fuse in the circuit can activate. This is an example of a fail-safe operation.

Battery system monitoring and control is required to ensure a correct and adequate response if an energy storage element fails. While this monitoring and control is required in all operating and non-operating conditions, it is especially important for charging when the battery accepts energy at elevated voltage. This control system must detect a failure and mitigate that failure in an appropriate time, to prevent the compounding effects of that failure. Control can be extended to passive portions of the battery, including thermal and structural elements, to mitigate failure effects.

For most EVs, the propulsion system is in a specific mode for charging; its HV bus is active although the propulsion system is depowered. Many EV architectures will use a common DC bus for the battery and the charging circuit, meaning isolation management of the complete HV system is necessary, when the vehicle is charging. Additional consideration is taken for access to HV, as the vehicle is stationary during the charging event. Many vehicles will also outwardly indicate their charge status, through fuel gages and/or activity lights that are visible when plugged in and the charging system is active.

Types of Charging

EV charging can differ markedly, depending on the type of device, location, and expected charging outcome. In many instances, the charging performance criteria have been standardized, so that solutions are provided that comply with a base set of specifications. For EV charging, one of the most ubiquitous standards is SAE J1772, which defines different levels of charge performance, for both AC and DC charging, as well as provides a specification for the manual interfacing of the EVSE. It operates in concert with IEC 62196. The focus of the following sections will be on the SAE J1772 standard of charging.

(A) Local Charging

Local charging covers the charging levels that can be provided by local AC infrastructure service. There are two levels defined under SAE J1772 that support this type of charging (Table 4.3).

TABLE 4.3 SAE J1772 AC-Level Charging Ratings

Charge Method	Voltage (V)	Phase	Max. Current, Continuous (A)	Branch Circuit Breaker Rating (A)	Max. Power (kW)
AC Level 1	120	1	12	15	1.44
			16	20	1.92
AC Level 2	208 or 240	1	24–80	30–100	5.0–19.2

AC Level 1 will normally be supported through a standard (120 V, 12 A) outlet, shown in Figure 4.6, while AC Level 2 will work with an oven (240 V, 32 A) outlet, shown in Figure 4.7. Both can be supported by OEMs using a convenience charging cable.

A standard pin arrangement and assignment are provided for both AC Level 1 and AC Level 2, as shown in Table 4.4. There are five pins in a standard arrangement. Line 1 (L1) and neutral (N) supply the main power link for AC service, while protective earth (PE) provides a ground return. The proximity pilot (PP) communicates between the EVSE and the vehicle that the vehicle is mechanically engaged with the EVSE, while the control pilot (CP) provides the charge control signal that is managed between the vehicle and the EVSE.

What is common in the AC L1 and L2 charging arrangements is that AC power is delivered to the vehicle and the vehicle contains the OBC, which converts the AC to

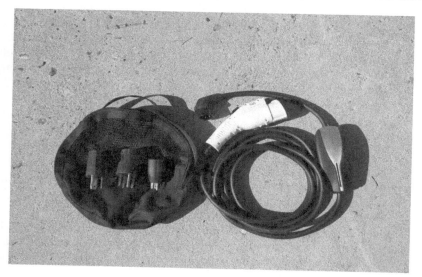

FIGURE 4.6 SAE AC L1 + L2 convenience charging chord. *Source: Paul Gipe/Wikimedia Commons/Public domain.*

FIGURE 4.7 SAE AC L2 wall-mounted EVSE. *Source: Igor/Adobe Stock Photos.*

How Infrastructure Drives Battery and Vehicle Design   101

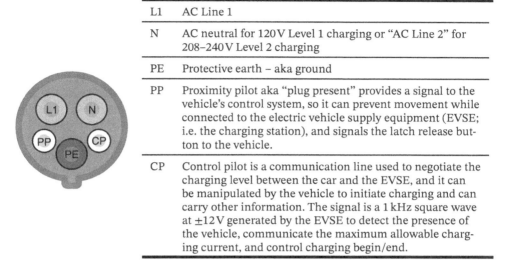

TABLE 4.4 Pin Arrangement and Descriptions for SAE J1772 Connector

L1	AC Line 1
N	AC neutral for 120 V Level 1 charging or "AC Line 2" for 208–240 V Level 2 charging
PE	Protective earth – aka ground
PP	Proximity pilot aka "plug present" provides a signal to the vehicle's control system, so it can prevent movement while connected to the electric vehicle supply equipment (EVSE; i.e. the charging station), and signals the latch release button to the vehicle.
CP	Control pilot is a communication line used to negotiate the charging level between the car and the EVSE, and it can be manipulated by the vehicle to initiate charging and can carry other information. The signal is a 1 kHz square wave at ±12 V generated by the EVSE to detect the presence of the vehicle, communicate the maximum allowable charging current, and control charging begin/end.

the appropriate DC voltage, to be supplied to the battery and vehicle auxiliary loads. The rate of charging will be dependent on the charge power capability of the OBC, since the charge rates associated with these levels are typically modest in comparison with the charge capability for most EVs.

AC L1 is a low-power charging, both in comparison with the charge capability of the battery and power electronics and with the power and load demands for the vehicle. Accurate vehicle power management is required when charging with AC L1. While the maximum charging power available can be as high as 1.92 kW, many convenience chord sets will be limited to closer to 1.2 kW continuous power. Additionally, the vehicle that is being charged may have a series of housekeeping operations that are preferred to run during charging. This could include items associated with the vehicle thermal systems and heating, ventilation, and air-condition (HVAC) or other on-board sensor suites. The cumulative loads could greatly reduce the charge that is delivered to the battery, or even discharge the battery, if not properly managed. For these reasons, AC L1 charging is considered for all, but the smallest EVs are used for backup purposes, although the accessibility of such power outlets warrants the vehicle maker to optimize the charging experience.

AC L2 has become ubiquitous in EV charging. It utilizes readily available 240 V and 32 A electrical service, as well as other similar services. The EVSE for AC L2 can be found in convenience charger chord sets, wall-mountable plug-in units, and permanently mounted wall units or in-ground posts. Like AC L1, the vehicle must supply the OBC that converts the AC supply power to the appropriate DC voltage. Much progress has been made over the past decade on OBC technology and design. The prohibitive costs of the power electronics had limited early OBC design to closer to a 16 A continuous current limit, meaning early OBCs were often limited to no more than 3.3 kW of charge power. Doubling the power electronic elements and optimizing packaging permitted the most common OBC charge level to rise to 6.6 kW. Today, many OBC designs readily accept up to 7.2 kW of continuous charge power. Charge conversion efficiencies have also improved and are commonly greater than 95% for most charge events.

The AC L2 charging standard will also support multi-phase charging, further increasing charging power. This charging arrangement is popular in certain regions of Europe and not readily available in North America. It would require the use of an IEC 62196 Type 2 (Type 2) connector and the appropriate OBC.

While local charging powers remain modest, even the 7.2 kW charge power for the AC L2 can be greater than what can be accepted by a cold battery in a PHEV. The vehicle will therefore need to manage the charge power of the OBC during the charge, which in turn communicates with the EVSE. The charge control is also required at the top of the charge for every charging event. At the top of the charge, the battery system will have limited charge acceptance, usually dependent on the current carrying limit of the cell balancing circuit and its associated charge balancing algorithm. The entire system control loop needs to be sufficiently accurate and nimble to respond to the balance control commands from the battery. The battery engineer will therefore include a voltage hysteresis allowance in their design that allows for this variation between the battery and the OBC. The OBC must also communicate at the appropriate rate with the EVSE to support this response, as well as compensate for the excess energy delivered or received between the EVSE and the battery, due to any inaccuracies beyond those tolerated by the battery charge algorithm.

(B) DC Fast Charging

Battery charging is a relatively lengthy process when compared to the fueling of an ICE-powered vehicle. Quicker battery charging is desired, and DCFC technologies, methods, and infrastructure have been developed to accomplish this. Unlike local charging, which relies on an AC supply, DCFC supplies HV DC directly to the vehicle. This means that the EVSE serves the function of both infrastructure supply and AC-DC conversion. There are many different levels of DCFC, depending on both the EVSE used and the infrastructure service that can be supplied.

SAE J1772 also covers DCFC configurations (Table 4.5). There are two common levels defined for DCFC: DC Level 1, which has a maximum current limit of 80 A, and DC Level 2, which has a maximum current limit of 500 A.

The Charging Interface Initiative e.V. (CharIN) consortium has developed a more comprehensive set of specifications for EVSEs to be used for DCFC (Table 4.6).

The CharIN classification approach recognizes the harmonized specifications defined by J1772 and permits operating window expansion to the limits of J1772 and beyond. An example of an HPC350 EVSE operating window is shown in Figure 4.8. The operating window is defined by the supply current and voltage from the EVSE, within the operating point limits that precisely describe the EVSE.

TABLE 4.5 DC Charging Levels Defined by SAE J1772

Charge Method	Voltage (V)	Max. Current, Continuous (A)	Max. Power (kW)
DC Level 1	50–1,000	80	80
DC Level 2	50–1,000	500	400

TABLE 4.6 Different CharIN EVSE DC Fast Charge Classifications

Power Class	Voltage Range		Min Current*	Current Range	Power Range	Communication
DC5	200 V	500 V	1 A @ 500 V	≤10 A @ 500 V	≤5 kW	DIN 70121
DC10	200 V	500 V	1 A @ 500 V	≤20 A @ 500 V	≤10 kW	DIN 70121
DC20	200 V	500 V	1 A @ 500 V	≤40 A @ 500 V	≤20 kW	DIN 70121
FC50	200 V	500 V	1 A @ 500 V	≤100 A @ 500 V	≤50 kW	DIN 70121
HPC150	200 V	920 V	5 A @ 500 V 5 A @ 920 V	≤300 A @ 500 V ≤163 A @ 920 V	≤150 kW	DIN 70121 and ISO15118
HPC250	200 V	920 V	5 A @ 500 V 5 A @ 920 V	≤500 A @ 500 V ≤271 A @ 920 V	≤250 kW	DIN 70121 and ISO15118
HPC350	200 V	920 V	5 A @ 500 V 5 A @ 920 V	≤500 A @ 500 V ≤380 A @ 920 V	≤350 kW	DIN 70121 and ISO15118

DC = direct current, FC = fast charging, HPC = high-power charging.
*0A min current is a special case and should always be possible.

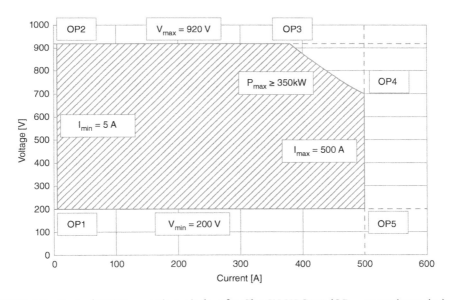

FIGURE 4.8 Typical EVSE operating window for CharIN HPC350 (OP = operating point).

A typical EVSE installation will provide the connector hardware for the vehicle, hard mounted to the host unit, shown in Figure 4.9.

The connector for DCFC has been developed in compliance with SAE J1772 and IEC 62196 and is known as the combined charging system (CCS), shown in Figure 4.10. In the case of the Type 1 connector, the changes were predominantly the

FIGURE 4.9 Example of a high-power DC fast charger station. *Source: Fastily/Wikimedia Commons/Public domain.*

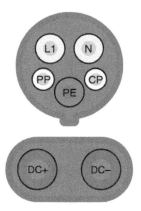

FIGURE 4.10 CCS Type 1 receptacle supporting DCFC.

addition of the DC+ and DC- pins below the main connector. In the case of the charger connector, L1 and N pins will be omitted, although the vehicle receptacle will retain these, so that both AC and DC charging can be supported.

The use of DC for the direct charging of the vehicle requires its own circuit within the vehicle, connecting the DC inlet contacts to the HV battery. This routing will typically bypass any OBC, although charging management and communication between the vehicle and EVSE will continue to be managed through the OBC.

(C) Vehicle Types and Considerations

The charging infrastructure is varied, and it needs to account for the specific needs and limitations of the different types of vehicles it interfaces with.

a. Passenger Vehicles

Light-duty passenger vehicles rely more on the accessibility and convenience of charging services than on almost any other form of mobility. Effective siting of EVSE stations for best use is key for the success of electrification of passenger vehicles. Installing an AC Level 2 charger within a single-family dwelling is simple, but this is not necessarily the case for multi-family dwellings. Similarly, charging at work is another opportunity, although there can be a challenge in providing enough charging capacity and stations by all employers.

As attractive as DCFC appears to be, it has considerable drawbacks in both the capital and operating costs, along with potentially enhanced degradation rates for an EV battery, when frequently fast charged. It is best reserved for longer-distance trips and is therefore expected to concentrate on long major roadways, interstate highways, and more remote sections of the country.

b. Commercial Fleets

EVs are often cited as ideal for use by commercial fleets, particularly those that are geographically limited. Such fleets often operate on controlled and fixed schedules, meaning that the scheduling for charging could be managed far more easily than for personally owned passenger vehicles.

Such captive fleets enable the owner of such fleets to make investments in their own charging infrastructure, with a rapid return on investments, because of their fixed and high-duty cycle. Depending on vehicle size, many commercial vehicles can make use of low-cost AC Level 2 charging, as opposed to relying on the installation of DCFC.

c. Heavy-Duty Vehicles

Heavy-duty vehicles will often require batteries considerably larger than those used by light-duty vehicles, because of their high vehicle demand energy (VDE). This is particularly the case when considering long-haul applications, where the battery sizes within the vehicles easily reach the 0.5–1 MWh range. At this size, DCFC is essential, with a newly emerging class of megawatt-level charging being considered essential for the adoption of heavy-duty EVs. Such stations will be highly impactful on the electric distribution grid, requiring infrastructure solutions unique to those supporting light-duty vehicles.

d. Bikes, Scooters, and Personal Mobility

Personal mobility devices are highly varied in design and use. In some instances, the battery is embedded within the device and in other cases removable. This means the charging hardware will be dependent on the device design. In almost all cases, the device will require a dedicated cable to connect to the infrastructure, usually a 120 V electrical port.

On board, the device will be the battery charging management system, which both supplies the DC power to the battery and manages the charging process, along with monitoring battery health. The design of such systems relies on the use of local and national electrical codes as well as standards created for the specific use of personal mobility devices.

(D) Low Temperatures

Batteries exhibit low charger performance at low temperatures because of their inherently sluggish chemical kinetics. This manifests itself in a higher internal resistance for the battery, leading to a greater voltage polarization on the battery when seeing a current delivered to or drawn from the battery. While charging, this means that the battery will achieve its upper voltage limit during charging sooner at low temperatures than it will at ambient temperature. Of greater concern in LIBs is that the anode potential within the cells may become sufficiently negative so that the lithium ions prefer to plate as lithium metal, rather than intercalate into the anode graphite. The lithium plating can lead to irreversible anode damage that can degrade battery life and potentially lead to cell failure and, in severe cases, thermal hazards through cell short circuits.

Reducing the charging performance of the battery at low temperatures will lead to longer charging times. For a PHEV, even the 7.2 kW charge power for the AC L2 can be greater than what can be accepted by a cold battery. As a result of this diminished performance, vehicle integration techniques pay particular attention to the battery thermal system, allowing for battery preconditioning prior to charging. It is common to provide a thermal system that can draw energy away from the OBC and use it to preheat the battery. Once the battery temperature has risen to a minimum performance threshold temperature, the battery will begin to accept charge power from the OBC. Prudent use of battery conditioning can enhance battery performance and optimize charging time.

2. Designing for Charging Interoperability

While PEVs have been commercially available for over a decade, the charging infrastructure continues to evolve. Charging hardware and interoperability standards have seen significant advancements and form the foundation for the charging infrastructure deployment. The concept of interoperability is that any PEV seeking to use charging infrastructure can reliably interact and operate with the infrastructure and that the infrastructure is ready to support any PEV that seeks to use its services. There is additionally a set of performance expectations set forth by the vehicle OEMs and operators that will govern the charging experience.

Charging hardware varies across the globe, and most critical to the PEV hardware is the coupler hardware style. The coupler is the hardware unit that interfaces the EVSE with the PEV receptacle port and facilitates electric power transfer and communication between the EVSE and the PEV. The coupler hardware can be generally broken into those supporting 1-phase AC, 3-phase AC, and DC charging, across the regions of North America, the European Union, Japan, and China. All other global regions will adopt from these regions.

An overview of the receptacles for couplers is provided in Table 4.7.

In North America, the predominant standards covering the charging interface and couplers are:

SAE J1772 – SAE Electric Vehicle Conductive Charge Coupler.

SAE J3068 – Electric Vehicle Power Transfer System Using a Three-Phase Capable Coupler.

SAE J3400 – NACS Electric Vehicle Coupler.

The SAE J1772 Type 1 coupler and the CCS Combo 1 couplers have been described previously. For 3-Phase AC, the coupler is defined under SAE J3068 as a Type 2 arrangement and is harmonized with the IEC 62196.2 Type 2 coupler used in the European Union. It incorporates additional L2 and L3 powered pins to support 3-phase AC charging.

North America additionally supports the North American Charging Standard (NACS) defined under SAE J3400. The coupler was originally developed by Tesla and

TABLE 4.7 Receptacles for Most Used EVSE Coupler Types

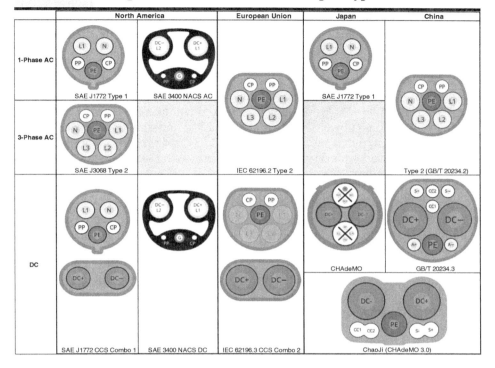

is now being utilized by other automotive OEMs. It can support both 1-Phase AC and DC charging, although not simultaneously, using common powered pins. For DC charging, it becomes an alternative to the CCS Combo 1 arrangement, allowing the use of the Tesla Supercharger network.

PRACTICAL INSIGHTS | SAEJ3400 and Interoperability in the Charging Network

Tesla Motors has been a unique company for a multitude of reason, not the least of which is that it has built its own infrastructure. A leader in scaled design and production of EVs, Tesla has developed, owned, and operated the largest charging network in the world. Until 2023, this network was for Tesla vehicles. Other vehicle manufacturers used connectors referenced elsewhere in this chapter, CCS (SAEJ1772) and ChAdeMO. That year, several large automakers, including General Motors, Mercedes, and Stellantis, announced they were adopting Tesla's connector to its proprietary network, known as the North American Charging Standard (NACS). This created a pathway to increased interoperability and for non-Telsa vehicles to take advantage of the company's large, well-established network. The Joint Office of Energy and Transportation requested the rapid standardization of NACS so that it could be cited in Federal Regulations for the National Electric Vehicle Infrastructure (NEVI), funded by the U.S. Federal Government as part of the Bipartisan Infrastructure Law which made $7.5 billion U.S. dollars available for an EV infrastructure. In response, SAEJ3400 was developed on a voluntary basis by engineers from across the supply chain.

The European Union uses a similar harmonized approach, governed through IEC 62196.2, for AC charging, and 62196.3, for DC charging. As mentioned previously, the 62196.2 Type 2 coupler can support both 1-Phase and 3-Phase AC charging. The CCS Combo 2 coupler will normally have its AC power pins depopulated, but otherwise operate identical to the CCS Combo 1 coupler.

Japan has historically used the J1772 Type 1 coupler for 1-Phase AC but utilized a domestically developed DC charging system, known as Charge de Move (CHAdeMO). It is a notably more complex coupling scheme than that defined by J1772/IEC 62196, but enables advanced communications and controls functions, as well as bidirectional power flow. The standard has undergone various revisions, CHAdeMO 2.0 being rated to 500 A and 1 kVDC. It had seen some popularity in regions outside of Japan but has generally dropped in popularity to the CCS standards, except in China.

China originally pursued a system based on IEC 62196, but with some alterations. For AC charging, the GB/T 20234.2 Standard uses a similar pin arrangement although the vehicle-side pins protrude, and the coupler becomes the receptacle. A complete native DC charging receptacle was defined under GB/T 20234.3. Both China and Japan are intend to move to a new, ultra-high power coupler arrangement with ChaoJi/CHAdeMO 3.0, which will permit powers of up to 900 kW.

Communications between the EVSE and the PEV are key to ensuring acceptable and reliable power transfer performance. For charging arrangements complying with J1772/IEC 62196, there are two pins that manage the signaling between the EVSE and the PEV:

(a) Control Pilot Functioning (CP), governed by IEC 61851-1 Annex A.
(b) Proximity Detection (AC and DC charging) and Current Coding (for DC charging) (PP), governed by IEC 61851–1 Annex B.

The control pilot is responsible for basic safety functions via a PWM signal as well as high-level communication (HLC), via a high-frequency power line communications (PLCs) protocol, superimposed on the pulse wave modulated (PWM) signal.

While most of the interest in the charging infrastructure has been associated with the charging of PEVs, there is the emerging case for the PEV also delivering power, as part of a distributed energy resources (DERs) strategy. Adding this interoperable functionality adds complexity to the entire vehicle-infrastructure ecosystem. To that end, a comprehensive communications, interoperability, and security management strategy has been developed and is governed under SAE J2836 and SAE J2931/7. The elements of the strategy are illustrated in Figure 4.11.

In addition to J2931, requirements and validation for vehicle-infrastructure interoperability are captured in SAE J2953. This ensures seamless and unambiguous verification of charging performance and function. Charging performance and EV charge rate definition are also being introduced to J2953/4, along with the joint standard ISO/SAE 12906. The intent of these standards will be to determine the EV charge rate performance in a definitive and unambiguous manner. The additional benefit of these standards is to connect UBE to both the vehicle and the charger performance.

Closely related to this ecosystem is "Plug-and-Charge," which strives to implement the appropriate communications protocols with the infrastructure so that a user's transactions with an EVSE do not need to be tied to a specific service provider.

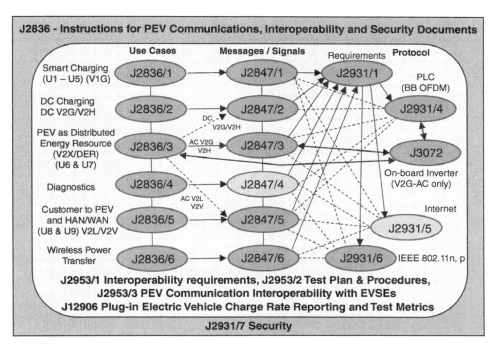

FIGURE 4.11 The J2836 ecosystem. *Source: with permission of SAE International.*

A summary of relevant standards for DER and bidirectional charging is provided in Table 4.8. These standards define the framework and execution for the transfer of energy to and from the PEV, as well as the communications protocols between the PEV and EVSE. Additional standards will exist for grid-specific power management, as well as for EVSE equipment and installations. ISO 15118 – Road vehicles – Vehicle-to-grid communication interface is the central and emerging global standard for vehicle-to-grid and other power export operations, generically known as V2X.

TABLE 4.8 List of Standards Applicable to Grid-Tied Vehicle Charging

Standard	Description
DIN 70121	DC charging
ISO 15118-2	All charging services as defined by Supported Energy Transfer Mode in subclause 8.5.2.3.
SAE J2847/2 (ISO 15118-2)	DC vehicle-to-home – V2H (EVSE inverter is grid forming). DC charging
	DC vehicle-to-grid – V2G (EVSE inverter is grid following). DC charging
SAE J2847/2 (ISO 15118-20 Light)	DC charging with BPT, physical layer according to ISO 15118-3 and ISO 15118-8
SAE J2847/3	IEEE 2030.5 – DER function set
SAE J2847/6	WPT, physical layer according to ISO 15118-8

(continued)

TABLE 4.8 Continued

Standard	Description
ISO 15118-20	Common messages
	AC energy transfer, physical layer according to ISO 15118-3 and ISO 15118-8
	AC charging with BPT, physical layer according to ISO 15118-3 and ISO 15118-8
	AC charging with DER, physical layer according to ISO 15118-3 and ISO 15118-8
	DC energy transfer, physical layer according to ISO 15118-3 and ISO 15118-8
	DC charging with BPT, physical layer according to ISO 15118-3 and ISO 15118-8
	DC charging with ACDP, physical layer according to ISO 15118-8
	DC charging with ACDP and BPT, physical layer according to ISO 15118-8
	MCS
	WPT, physical layer according to ISO 15118-8
ISO 15118-200	Extensible SECC Discovery Protocol and Error Information Protocol

3. How New Infrastructure Affects the Supply Chain

The new mobility electrification infrastructure creates a series of vehicle-side and EVSE-side impacts, which are felt within the entire supply chain. Vehicle OEMs both influence and are influenced by the infrastructure and EVSEs are subject to the simultaneous demands and constraints of both the vehicles and the infrastructure.

The current infrastructure environment for mobility is described in Figure 4.12.

Most mobility devices, including passenger vehicles and public transit, can be considered to be operating locally. For personal mobility, it can be broken into localization of the mobility device to a single-family dwelling or a multi-family dwelling.

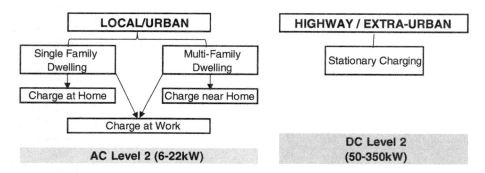

FIGURE 4.12 Example for the current charging infrastructure ecosystem for passenger vehicles and mobility devices.

One of the major purposes of mobility is for commuting so charge events can occur at either point in the trip, when the vehicle is stationary. Small mobility devices, such as scooters and e-bikes, offer a degree of portability. Larger mobility devices are limited and must be parked at some location, so this limitation becomes critical to siting charging infrastructure. The most popular charging infrastructure is AC Level 2, which can be conveniently installed in single-family dwellings, workplaces (Charge-At-Work), and places of commerce. Multi-family dwellings face unique siting challenges, and this is a significant consideration for infrastructure planning and management. Associated with this would be street-side charging infrastructure, which can support a wider range of mobility solutions.

Long-distance travel and extra-urban trips will require a different infrastructure than that used locally. The current focus remains on stationary charging infrastructure, mimicking that of conventional fueling stations. It is that analog which has driven increased charging performance and particularly the development and deployment of DCFC. This infrastructure is substantial in its content and installation, and significantly impacts the local electrical grid infrastructure. It is common to match DC fast charge stations with additional stationary battery energy storage systems (ESS) to provide an energy buffer that mitigates stress on the hosting grid service, as well as minimizing the resultant demand charges that would otherwise be incurred by the charging infrastructure provider.

There are therefore two key aspects influencing the selection of the appropriate charging solution for a vehicle: performance and availability.

Performance improvement is summarized as the time to charge the vehicle or device. The value can mean different things to different users. For example, a small mobility device, such as an e-bike, may be satisfied with time to recharge the battery, while an EV passenger vehicle may be more interested in miles per minute of charge. Furthermore, a battery supplier may be interested in the time to return a specific amount of UBE within a specific SOC range. A good example of charge rate reporting is that described by SAE J2953/4 – Plug-in Electric Vehicle Charge Rate Reporting and Test Procedure. There are physical and electrochemical considerations that govern performance. Charging power is physically governed by both the electrical voltage and the current of the complete system (vehicle, EVSE, and infrastructure), which are in turn governed by the limitations of each of the components within the system. The battery itself will utilize a particular electrochemistry and cell design, which determines the rate and efficiency by which electric charge can be accepted.

Electric current is one of the single greatest enablers and limiters to charging. Most of the electrical hardware is designed in accordance with the selected current limits. Electric currents are normally not an issue until DCFC for PEVs, where current limits as high as 500 A are targeted. Such currents create thermal challenges for all components involved and pose unique stresses on devices such as thermal fuses and connector seals. For the EVSE the current limit has a particular impact on the sizing of conductive cables, as well as the need for introducing liquid cooling with some cable designs. All upgrades in the current limit impact materials selection as well, for example, utilizing considerably more copper in components where aluminum might have been an alternative.

System voltage also influences charge power, with higher voltages having the benefit of either increasing charge power or lowering charge currents. Lower charge currents reduce system losses from resistive heating, increasing overall operational efficiency. Maintaining the same charge current and increasing voltage multiply the supply power. It is this method that was employed in increasing DC fast charge from

150 to 350 kW. Higher voltages do require the vehicle-side and supply-side equipment to meet higher-voltage creepages and clearances, which in turn do drive a modest amount of additional cost.

Charging reliability of performance is another significant consideration. The vehicle and the EVSE may have different performance capabilities, based on the limits of current and voltage. Interoperability standards are expected to allow for operation between the EVSE and the vehicle, but ensuring optimal charging performance and delivery of charge to the vehicle can be a challenge. On the vehicle-side FC will be managed as a function of the battery SOC, measured voltage, and measured temperature. Additionally, advanced algorithms will account for battery age, or state-of-health (SOH), as well as distinct electrochemical phenomena, such as lithium plating potentials. Battery charging profiles can therefore be dynamic and complex, meaning the real charging time for a particular event can vary relative to previous events. Furthermore, the communication with the EVSE that is intended to control the charge demand current and voltage may exceed the EVSE's ability to respond, meaning the EVSE can lower charging performance or stop altogether, to ensure the safe operation of the charge event.

Critical to infrastructure is the availability of service. As indicated above, there are particular types of EVSE that are preferred for particular use cases, and, within those types, there can be a diversity of performance ranges and features. At the start of 2024, there were approximately 61,000 publicly accessible EVSEs, of which 28,000 were considered fast chargers (U.S. Department of Energy – Energy Efficiency and Renewable Energy 2023). This number can be compared to approximately 111,000 gasoline refueling stations that currently exist. While the number of charging locations increases the duration of a charge event means that a vehicle will occupy a station considerably longer than it would at a conventional refilling station, meaning that there will be a greater relative need for charging stations than for refilling stations.

Case Study: ChargerHelp, EVSE, and the Path to Cleantech Jobs

Los Angeles-based ChargerHelp has a mission to remove barriers so that people in disadvantaged communities have access to opportunities in the cleantech, electronic equipment repair industry and to promote economic mobility by providing good paying jobs. ChargerHelp has partnered with workforce development agencies throughout the United States to recruit candidates from all skilled backgrounds and train them to become industry experts as EVSE technicians and to help solve the issues that EV charging station owners and EV drivers experience on a daily basis.

By utilizing technology, ChargerHelp has enabled trained local EVSE technicians to identify core issues, capture data, and centralize it to help solve those issues. In partnership with charging companies, ChargerHelp has surveyed the functionality of over 20,000 EV charging stations in a multitude of environments and climates. ChargerHelp has found that the vast majority of concerns reported is the absence of consistent maintenance plans and reliability challenges across different communities. For this reason, EVSE manufacturers, network providers, and utility companies trust ChargerHelp to deliver practical solutions that will support the evolution of vehicle charging technology.

ChargerHelp, with the support of safety training centers, has identified three fundamental training requirements that any person interested in operating and maintaining Level 1 and Level 2 EV charging stations must complete as first steps to becoming an EVSE technician. These trainings include in-person or online OSHA 10-Hours for General Industry, including Lock Out Tag Out (LOTO), and NFPA 70 E for General Industry. These training courses could all be completed in less than one week.

If an EVSE technician is interested in operating more complex equipment like Direct Current Fast Chargers (DCFC) or similar, in addition to the three fundamental trainings listed above, OSHA 30-Hours for General Industry and the High-Voltage Safety Training are the most appropriate. Though an EVSE technician's work does not involve electrical work, these courses are administered to limit any unintentional exposure to electrical and chemical hazards. Incidental electrical risk training courses could all be completed in one week.

Once the fundamental safety training is completed, candidates may take an EVSE operations and maintenance (O&M) training through an EV charging station's manufacturer and EV charging software courses from EVSE network providers. Candidates may also receive training with a company like ChargerHelp, who teaches a robust curriculum on a wide range of parts and software complexities such as Open Charge Point Protocol (OCPP),[2] Open Charge Point Interface (OCPI),[3] and the International Organization for Standardization (ISO).[4]

Because the basis of operating internal electronic components is easy to learn and is low risk, it is imperative that this line of work remains open to the public, limits additional prerequisites, and limits creating bottlenecks for people in low-income, disadvantaged communities, or a community of people transitioning from industries like telecommunications, nuclear, coal, oil, gas, and even traditional automotive mechanics into the clean energy industry.

Charge event reliability is another area of concern, as this has been a challenge to achieve. In 2023, the U.S. Department of Transportation made a $100 million investment through the Bipartisan Infrastructure Law, to improve charger reliability, with a goal of exceeding 95% reliability in achieving a successful charge event (U.S. Department of Transportation 2023). The charging infrastructure has experienced a wide range of issues that have prevented it from approaching the reliability goal, estimated today to be closer to 78% (DeLollis and Justice 2024).

Reliability issues have been captured and classified and many of the common issues are summarized in Table 4.9. The largest issue is the charger being unavailable, or un-powered. The inability to perform a payment transaction is the second most common issue. Connector and communication errors are the next most common issues.

For a charging customer, the particularly notable issues are those related to payment and communicating with the charger. The frequency of these failures, along with the commercial issues related to having a wide and disparate array of charging service providers, has led to the Plug-and-Charge concept. Plug-and-Charge is a technology designed to streamline the process of charging EVs by automating the

---

[2] https://www.openchargealliance.org/protocols/ocpp-201/
[3] https://evroaming.org
[4] https://www.iso.org/sites/outage/

TABLE 4.9 Most Common Reasons for Charge Event Failure

Reliability Issue	Percentage (%)
Charger unavailable	25
Payment system failure	20
Connector malfunction	15
Software/communication error	10
User error	10
Power supply issue	8
Other	12

authentication and payment process. When an EV equipped with Plug-and-Charge technology is connected to a compatible charging station, the vehicle and the charger communicate directly to authenticate the user and initiate the charging session. This eliminates the need for drivers to use RFID cards, mobile apps, or credit cards to start the charging process, making it as simple as plugging in the vehicle.

The primary benefit of Plug-and-Charge is the enhanced convenience and user experience it offers. By reducing the steps required to begin charging, it minimizes potential points of failure and speeds up the process. This technology also enhances security, as the authentication is handled through encrypted communication between the vehicle and the charger. Overall, Plug-and-Charge aims to make EV charging as seamless and hassle-free as possible, encouraging more drivers to adopt EVs.

Plug-and-Charge is expected to become commonplace among the charging infrastructure and is expected to be forward compatible with new charging technologies and features.

4. How the Electrification Movement Affects the OEM's Technology Roadmaps as They Migrate from ICEs to BEVs

The electrification infrastructure can be perceived in a broad manner, where the PEV is a source of energy as well as used, distributed among many other sources and users, coupled together by a flexible and adaptable charging infrastructure. While many of the key elements enabling such an infrastructure have been defined and are being implemented, there is a continuous demand for growth and diversification within this ecosystem. This relationship between the vehicle and infrastructure is distinctly different from that experienced previously with ICE vehicles and has significantly impacted the technology roadmaps for automotive original equipment manufacturers (OEMs). The OEM now considers the implications of vehicle use for functions other than simple mobility. Additionally, since electricity can be transmitted in multiple manners, there can be different technologies employed for different purposes.

Electric energy can be transferred to a vehicle wirelessly, in place of conventional conductive charging. Inductive charging has been used for a sometime, in contrast with the conductive charging that is predominantly used today. It was used in the GM EV1 for its charger, supplied by Delco Electronics, as the Magne Charge inductive charging system. It was selected primarily for the heightened level of safety

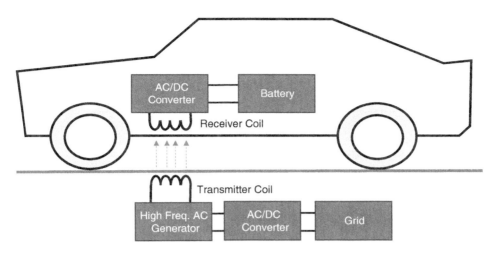

FIGURE 4.13 Schematic representation of a wireless charging network.

perceived when using an inductive charge transfer approach. Wireless charging now assumes the vehicle is completely independent of the charger hardware. The dominant mechanism used for charging is resonant inductive coupling, where the charger is an inductive pad that requires alignment with an element on the vehicle to create the inductive coupler and transfer the electrical energy, shown in Figure 4.13.

The benefits of wireless charging include:

(a) No direct contact with DC within the charge transfer system, meaning a significant reduction in shock risk.
(b) No mechanical connection between the EVSE and the PEV, requiring less direct interaction with the hardware equipment.
(c) A wide power range available, allowing versatility in installation.

Wireless charging has been developed to the point where it has been implemented with commercial and captive fleets, as well as for non-roadable applications. The predominant standard governing vehicle wireless charging is SAE J2954: Wireless Power Transfer for Light-Duty Plug-In/Electric Vehicles and Alignment Methodology (Table 4.10). It was originally developed to cover charging levels that were analogous to J1772 but has now expanded to include specifications for wireless charging up to 500 kW, in J2954/2.

The operation of a WPT pad requires the vehicle to be sufficiently aligned with the transmission pad so that the receiver pad on the vehicle can power couple with the transmission pad. The alignment strategy is also provided within J2954, known as the minimum-common alignment method, and enables user-friendly interfacing and communication with the vehicle operator, to guide the vehicle to the correct location. Like auto parking, this feature can also be automated.

DWPT is a variation on static wireless power transfer, when the vehicle can be continuously moving over a train of transmission coils embedded in the roadway. Sometimes known as a Wireless Electric Road Systems (wERS), it provides a function like a wired system, such as those used for urban transit and rail, but with no conductive electric power transfer. Power transfer can be significant with most systems under study being in the WPT6–9 range, with the intent to power electric commercial

TABLE 4.10 Wireless Power Transfer (WPT) Levels

WPT Level	Power (kW)
1	3.7
2	7.7
3	11
4	20
5	50
6	75
7	150
8	250
9	500

vehicles. There are many pilot programs currently underway in the United States, Germany, France, and Italy, with many more planned. In the United States, Utah State University led NSF Engineering Research Center, known as ASPIRE, leads research into a totally integrated infrastructure based around DWPT roadways.

While conductive charging infrastructure has been shown to have some considerable concerns over their installation, inductive charging systems have their own unique considerations. Such WPT stations are costly to install given the degree of roadwork involved with the hardware and controls installation. Concerns remain over the effect of road debris and precipitation, water, and snow on the pad's performance, availability, and safety. Road maintenance would now also need to account for the WPT hardware, during road repair and upgrades.

An EV can be envisioned as a mobile energy source, which is able to export electrical power. The concept of vehicle electrical energy export is often denoted as vehicle-to-demand (V2X), suggesting the demand can be any source, such as another vehicle, a residence or business, accessory power demand, or the electric grid. As such a vehicle can now be envisioned as being part of a much broader energy ecosystem. This was described above as being under the governance of SAE J2847 and ISO 15118.

For the EV employing V2X functions, it must be capable of bidirectional power transfer (BPT). In the case of DC charging, this is a simple reverse power flow from the battery to the DCFC EVSE. In the case of AC charging, the charger itself must be capable of converting battery DC power to the appropriate AC power, in terms of voltage, current, and phase, so that it can be exported (wired or wireless) to the AC EVSE.

The infrastructure hardware will depend on the V2X role to be performed (Table 4.11). Vehicle-to-Load (V2L) is where the vehicle powers external loads, and where those loads are self-regulating. Today, many vehicles have on-board inverters that supply nominal AC power for accessory devices. The concept can be expanded to utilize the power capability for the powertrain on-board inverter, or independently tied to the vehicle HV propulsion system bus. This way it is possible to power a wide range of external devices.

TABLE 4.11 Different Examples of Vehicle Power/Energy Demands

V2L	Vehicle-to-Load	Load does not communicate to vehicle, is regulated by vehicle
V2V	Vehicle-to-Vehicle	Transfer between vehicle, communication is managed
V2H	Vehicle-to-Home	Transfer to home, vehicle regulated by home load
V2B	Vehicle-to-Business	Similar to home
V2G	Vehicle-to-Grid	Transfer to grid, vehicle regulated by grid service operator

Vehicle-to-Vehicle (V2V) utilizes the vehicle on-board communication to transfer power to another vehicle. Only the supply vehicle needs to have the ability for bidirectional charging, and the coupling cable will generate and manage the communication between both vehicles. It is this way possible to manage the transfer of both AC and DC power, depending on the selected cable's configuration. The rate of power transfer will be restricted by the capabilities of both vehicles and the current limit of the cable hardware.

Vehicle-to-Home (V2H) and Vehicle-to-Business (V2B) enable a vehicle to supply power to the building, via a bidirectional charging capable EVSE. The power may be used for power backup & reliability, load-shifting, solar or other renewables firming, and power supplanting. These are considered behind-the-meter operations, and do not follow the control demands for the electrical grid. The building therefore needs to provide isolation from the grid to the power net using V2H/V2B. The building will control the time of use and degree of use of the EV battery energy, tied with the restrictions in total energy and time selected by the vehicle user. Private businesses could incentivize the participation of vehicles in V2B through payment of service and/or discount of charging services at the business location.

Vehicle-to-Grid (V2G) ties the vehicle to the electrical grid and therefore operates in accordance with the demand requests from the grid. As such, the vehicle–EVSE communication is governed through a "grid-forming" protocol, where once the vehicle is connected and synchronized with the grid, it can perform bidirectional energy storage functions. In addition to the functions provided for V2H/V2B, the vehicle can support grid-level operations, such as voltage and frequency stabilization, grid firming, and power deferral (accepting excessive energy from the grid). The vehicle user will effectively "sign on" the vehicle to allow its use for a prescribed period. The grid will normally aggregate vehicles and other energy storage to meet demand requests from the ISO. Grid service providers refer to a vehicle using V2G as a distributed energy resource (DER). Many grid service providers and the respective aggregators will provide payment for this service, which can more than offset the cost of the battery charging, and in some instances by a significant margin.

The use of EVs in home and grid energy storage is attractive, particularly when considering the potential financial benefits. This does have an impact on the vehicle battery durability, however. While most modern LIBs have excellent cycle life and energy throughput, they are finite. A popular method for determining the impact of V2X on a battery is to monitor the amount of energy exported from the battery and converting to equivalent miles of vehicle use, as defined by the range certification process. The method has been defined in UN EVE GTR No. 22, as virtual distance (in km). The plot in Figure 4.14 illustrates the cumulative effect on vehicle mileage, when being used for V2X, in miles per year. For simplicity, it assumes a 4-hour daily

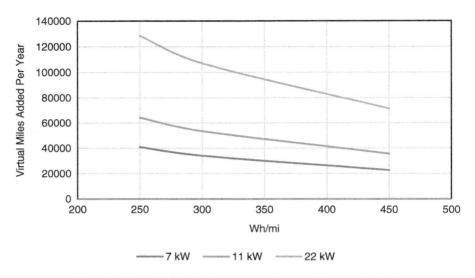

FIGURE 4.14 Virtual miles added by V2X assuming 4 hours daily use for discharge.

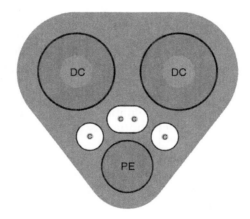

FIGURE 4.15 Proposed MWC coupler configuration.

usage window at a few examples of common charge rates. The figure shows that regular use at even low powers can rapidly accumulate virtual distance. For this reason, there has been some controversy over whether such very low usage powers will result in battery degradation to the use in vehicle operation.

Recently DCFC has advanced to the level of delivering as much as 400 kW charge power on a continuous basis. Still there are applications that would benefit from yet higher charge rates. For example, it is common for long-distance electric busses and Class 7/8 trucks to utilize upward of 600 kWh battery storage. Additionally, marine and aviation batteries are proliferating, with a particular focus on systems voltages into the 1,200 VDC range. This emerging need has brought the advent of the Megawatt Charging System (MCS), offering up to 3.75 MW of charge power. The MCS envisions electric currents up to 3,000 A and up to 1,250 VDC. The MCS connector/coupler is being proposed by the CharIN Organization, as shown in Figure 4.15.

The system follows a similar arrangement as CCS, but with expanded communication capabilities. The large size of this interface will propose some unique challenges in the handling of the connect/disconnect process, along with the additional thermal management required for the significantly increased currents. Such power demands will almost ensure that the systems are deployed with additional battery energy storage, to reduce the demands on the localized electric grid supplying these charging systems.

An alternative to charging the battery within the vehicle is battery swapping, or removable charging. This is commonly used in small electrical vehicles such as e-bikes and scooters and in lawn and garden applications. Battery swapping has also been adopted in a limited amount for light-duty EVs. In all cases, particular attention is paid to the electrical power and communication interfaces, given the repetitive nature of connect and disconnect operations. Additional care needs to be taken to design robustness against pollution, fouling, and abuse of mechanical loads. The pack structural characteristics need to account for the interfacing with the host system and will require connect/disconnect mechanisms that ensure positive engagement with the host body and not negatively impact the structural performance of the system. If the pack has a thermal system, air passages, and liquid coolant interfaces will also pose a distinct challenge to prevent fouling/leakage. An overview of pros and cons of swapping in light-duty EVs is provided in Table 4.12.

While swapping the battery in a personal mobility device can be done locally, light-duty EVs will rely on dedicated and specialized swap stations to perform the battery swap operations. Additionally, a supply of batteries is required for each swap station to support the respective vehicles. The swap station will also recharge a battery from a previously serviced vehicle, as well as have the option to assess battery state of health and share such information with a centralized battery inventory management system.

Battery swapping infrastructure can be fixed or mobile, allowing for differing degrees of scalability and agility. This is what contrasts one swapping strategy with another. Different strategies can extend to the battery itself. Swapping a single, complete battery is the simplest approach and will result in the shortest swap times.

TABLE 4.12 Battery Swapping Pros and Cons for Light-Duty EVs

	Pros	Cons
Charging	Faster than DC fast charging	Swap station needs sufficient battery charge capacity to handle demand
Range	Potentially scalable, depending on pack architecture. Range may no longer be an issue.	Lower range in most batteries, and relies on swap station availability
Integration	Existing vehicles can be retrofitted to swappable batteries	Higher weight, and complexity with interfaces. Each vehicle may demand unique hardware.
Cost	Charging costs lower than DCFC	Higher hardware costs plus swap stations

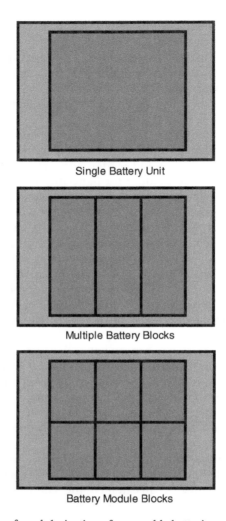

FIGURE 4.16 Examples of modularization of swappable batteries.

It is a challenge to standardize a single battery between different vehicles, given diverging requirements and platform sizes. A single OEM may therefore choose to utilize a block approach, where several smaller battery blocks may be carried within a single vehicle. The blocks are usually connected in parallel to a single HV bus, allowing for the option to select different vehicle ranges depending on the number of blocks installed, as well as support multiple platforms. This concept can be further extended to smaller modular blocks, which allow for a more diverse range of integrations. These concepts are depicted in Figure 4.16.

The benefits of using smaller blocks are the ability to separate some functions more easily, such as thermal management systems, from the modules. The small module size also means low relative currents per block, when compared to the total pack current. This means more sophisticated systems management tools such as DC/DC converters may be incorporated into the modules, allowing for mixing modules of different chemistries, states of health, and voltage. Battery

> **PRACTICAL INSIGHTS** | Nio, Inc., an Example of a Customized Infrastructure Solution
>
> Some automotive OEMs prefer to manage at least part of their own infrastructure rather than rely on a communized infrastructure approach. This is often to provide a differentiation in the products and services they provide, as well as attempting to evaluate alternative technologies and approaches. One OEM example like this is Nio, which intends to offer higher performing charging solutions than are available through standard infrastructure.
>
> Like many other OEMs, Nio offers a range of vehicles, based on a common core platform. All vehicles are EVs. Nio has its own EVSE and charging service provider, known as Nio Power, which provides charging options to individual customers as well as commercial service providers. It offers conventional AC Level 2 home charging, as well as DCFC of up to 660 A/550 kW, in support of upcoming ChaoJi standards. All vehicles that Nio offers would be able to be supported by these EVSEs.
>
> Nio also created a Battery as a Service (BaaS) business concept, which offers battery swapping to all owners of its vehicles. Nio has claimed that BaaS has reduced vehicle purchase price by about 25% (Elsayed et al. 2024). This is accomplished by offering the battery as a separate item to be leased, as a service from Nio. Key to the service is to offer different batteries with different energies that suit the range needs of its customers. All batteries are single units which can be swapped at Nio's stations or simply used in the vehicle in a conventional manner.
>
> The three batteries that are standard in the current fleet are:
>
> 1. 70–75 kWh "Standard Range" LFP pack offering ~500 km NEDC range
> 2. 100 kWh "High Range" NMC pack offering ~700 km NEDC range
> 3. 150 kWh "Ultra High range" hybrid solid-state pack offering ~1000 km NEDC range
>
> Lease costs for each battery will vary, but this approach also allows the introduction of newer technologies in a more captive manner. Nio has demonstrated this with the launch of their hybrid solid-state battery.
>
> The battery swapping station network has over 1,300 battery swap stations in China, targeted primarily to expressways, as an alternative to DCFC. Nio has also recently expanded into Europe with approximately 43 charging stations across Scandinavia and Germany with plans to bring the system to the United Kingdom and potentially the United States. Nio is also offering swapping for other OEMs to expand standards associated with battery designs, swapping technology, asset management, and services. Some other Chinese OEMs have accepted the offer.
>
> Source: NIO *(20 August 2020)*. "NIO Inc. Announces Launch of Battery as a Service and Establishment of Battery Asset Company." Retrieved 17 October 2021 – via **globenewswire.com**.

swap time is greater for most systems employing multiple blocks, than for a single battery unit.

Battery swapping creates an interesting challenge for battery life and durability. The customer can be masked from the loss of performance of a battery, although they will receive random batteries with a wide range of SOH, and variable resultant vehicle range. Such variations in the batteries can be reduced assuming the vehicle can communicate with the swap station and have that station provide the battery of range sufficient to finish the journey or reach another swap station along the route. The battery swapping service provider may still be required to provide accurate SOH determination for the batteries, when they are installed in the customer's vehicle. The battery swapping service provider also has the opportunity to evaluate battery technologies that may otherwise pose a challenge for durability, while still satisfying the customer's range expectation.

> ### Future Considerations
>
> EVs and EV infrastructure constitute a dynamic area. Their development can be affected by economic and regulatory changes. The International Energy Agency (IEA) indicated that 14 million EVs were sold in 2023, with 95% of sales in China, Europe, and the United States (**https://www.iea.org/reports/global-ev-outlook-2024/trends-in-electric-cars**). Notwithstanding, several factors are worth noting for future consideration with respect to infrastructure.
>
> - EV charging reliability, while improving, is still sub-par when compared to the reliability of the fueling infrastructure of ICE vehicle.
> - Although the standardization of Tesla's NACS represents a major step forward in charging infrastructure reliability, access to charging stations, particularly in rural or isolated areas, may remain a drag on consumer confidence in EVs.
> - As of 2024, fewer than ten (10) NEVI-funded charging stations had been installed in the U.S. signaling that while public charging is robustly funded in the United States, the establishment of a widely available charging infrastructure will take time.

Words to Know

Bidirectional charging   The use of an EV battery as a storage source for extra-vehicular sources.
EVSE   The general term for the infrastructure that supports the charging of the EV, such as charging stations.
Gravimetric energy density   Also known as specific energy; the discharge energy capacity of the battery divided by its mass under specified discharge conditions.
Hybrid electric vehicle (HEV)   An electrified vehicle where traction is supplied by more than one power source.
Plug-and-Charge   A service and process where the vehicle and infrastructure wirelessly communicate to manage the charging process and all data and financial transactions without any inputs from the user.
Plug-In Electric Vehicle (PEV)   An electrified vehicle that plugs into the infrastructure to receive some or all of its traction energy. This includes Battery Electric Vehicles (BEV), Plug-in Hybrid Electric Vehicles (PHEV), and Range-Extended Electric Vehicles (REEV).

Solid Electrolyte Interphase (SEI) Layer   The electrically resistive and ionically conductive protective layer that forms on the anode of a lithium or lithium-ion cell that allows the cell to cycle.

Stranded Energy   Energy remaining in a battery that has been electrically and/or physically disconnected from the vehicle and cannot be accessed, normally after a crash incident.

Thermal Runaway   An uncontrolled increase of battery temperature caused by exothermic reactions inside the battery.

VDE   The energy required to provide traction to the vehicle under the set operating conditions of that vehicle.

Further Reading

EV charging infrastructure is a dynamic and fast-evolving field. The journal literature in this area changes rapidly and we recommend consulting the major publishers and aggregators of science and technical literature in this area, namely Elsevier's ScienceDirect (**https://www.sciencedirect.com/https://www.sciencedirect.com/**) and Wiley's WileyPLUS (**https://www.wileyplus.com/platforms/wileyplus/**)

The University of California, Davis's Electric Vehicle Research Center with the Institute of Transportation Studies also publishes pertinent information on charging infrastructure. **https://ev.ucdavis.edu/product-area/charging**

References

Borja, A. (2024). Photo essay: e-bike fires leave scars on buildings across New York City. *City and State New York* (8 April).

Byington, L. (2022). Series of Tesla fires in Florida linked to saltwater damage from hurricane Ian. *Insurance Journal* (17 October).

Crowley, K. and Weise, E. (2023). Hurricane Idalia floodwaters cause Tesla to combust: what to know about flooded EV fires. *USA Today* (1 September).

Dekraker, P., Kargul, J., Moskalik, A. et al. (2017). Fleet-level modeling of real world factors influencing greenhouse gas emission simulation in ALPHA. *SAE International Journal of Fuels and Lubricants* 10 (1): 217–235. US Environmental Protection Agency, SAE 2017-01-0899.

DeLollis, B. and Justice, G. (2024). The state of EV charging in America: Harvard research shows chargers 78% reliable and pricing like the 'Wild West'. *Harvard Business School* (26 June). **www.hbs.edu**

Dutton, H. What Building Owners Should Know About the Rise of E-Bike Battery Fires (2024). *Propmodo* (13 February).

Editorial Team (2022). Lithium-ion batteries: Fire risks and loss prevention measures in shipping *Safety4Sea* (31 August).

Elsayed, M., Zafar, K., Esa, Y. et al. (2024). Impact of 100% vehicle electrification on the distribution grid in dense urban regions. *Energy Reports* 11: 5315–5322. Grid limitations to support fast charging, ScienceDirect. **https://doi.org/10.1016/j.egyr.2024.05.030**

Gelinas, N. (2023). Following fatal battery fires, NYC needs more rules for e-bikes. *New York Post* (24 June).

Glavan, S. and Burke, E. (2022). *Allianz: Prevention Measures Crucial to Tackling Risk of Battery Fires in Shipping*. Business Wire (31 August).

Impelli, M. (2023). Tesla erupts in flames after getting flooded during hurricane. *Newsweek* (31 August).

NTSB (2020). Safety risks to emergency responders from lithium-ion battery fires in electric vehicles. *Safety Report*. NTSB/SR-20/01, PB2020–101011.

Schuler, M. (2022). *gCaptain* (10 March).

Tefft, B. C. (2022). Research brief, American driving survey, 2020–2021. *AAA Foundation for Traffic Safety* (October).

U.S. Department of Energy – Energy Efficiency and Renewable Energy (2023). The information source for alternative fuels and advanced vehicles. Alternative Fuel Data Center. **www.afdc.energy.gov**

U.S. Department of Transportation (2023). Biden-Harris administration making $100 Million available to improve EV charger reliability. *U.S. Department of Transportation* (13 September). **www.transportation.gov** (accessed 21 June 2024).

CHAPTER 5

Lithium-Ion Battery Recycling

Steven Sloop
OnTo Technology LLC, Bend, OR, USA

Introduction

As lithium-ion battery (LIB) production for electric vehicles (EVs) and energy storage systems (ESSs) grows, recycling LIB scrap to make new batteries is crucial for economic sustainability. The automotive industry demands reuse, refurbishment, and materials recycling. While small-format battery manufacturing started in 1991 and is well established, large-format LIB production for EVs is still developing. Market growth is driven by volumetric energy density and flexibility of energy sources for charging, but hindered by costs, safety, and performance issues. Despite lower LIB costs, EV demand dipped in 2024. The industry aims to further improve LIB costs through innovation and recycling, aiming for materials circularity. Engineers and scientists must grasp the processes, challenges, and opportunities in LIB recycling.

What You Will Learn in This Chapter

This chapter covers chemical methods for recycling LIBs, including industrial case studies on pyrometallurgy, hydrometallurgy, and preindustrial direct recycling processes. Technical sidebars provide detailed chemical insights relevant to engineering and management in LIB recycling. The chapter describes types of battery scrap and pretreatment methodologies for safe recycling and concludes with a discussion on battery design for recycling.

In this chapter, readers will gain fundamental knowledge and comprehension of the following learning objectives:

- Knowledge of bill of materials (BOMs) present in LIBs:
  - Nickel, manganese, and cobalt.
  - Iron and mixed metal phosphate.

Electric Vehicle Batteries: From Sourcing to Second Life and Recycling, First Edition.
Edited by Bob Galyen and Frank Menchaca.
© 2025 John Wiley & Sons, Inc. Published 2025 by John Wiley & Sons, Inc.

- Knowledge of three technical recycling methods:
  - Pyrometallurgy
  - Hydrometallurgy
  - Direct
- Application of the knowledge to understand the economic value limitations/opportunities for each recycling method applied to the two general LIB BOMs.
- Understanding of the technical matching of recycling methods applied to the two general LIB BOMs with respect to carbon footprint opportunity and value opportunity.
- Knowledge of the pretreatments applied to LIBs for recycling:
  - Discharge
  - Neutralization
  - Black mass preparation to salts and precursor cathode active material (PCAM)
- Understand the logistical safety challenges and opportunities for end of life (EOL) LIBs.
- Knowledge of the two types of scrap materials:
  - New scrap
  - Old scrap
- Understand the recycling system designs necessary for each type of scrap material.
- Knowledge of the potential phase-out of polyfluorinated alkyl substances (PFAS).
- Understand the implications of PFAS and solvent phase-out for recycling as a function of the recycling types.
- Knowledge of the hierarchy of approaches in design for recycling:
  - Cell chemistry and packaging
  - Battery and module packaging
  - Battery management systems.

End-of-Life Definition

A battery hits EOL when its potential value of usable deliverable energy drops below the value of its internal chemicals (B_{sv}). However, in practice, EOL is triggered if it is impractical to relocate or service the battery for another lower-performance application.

To evaluate this concept, we approximate the energy delivered over a battery's lifetime to be E_T(Wh/kg) $= N_i * E_D/1000$, where N_i is the number of cycles and E_D is the specific energy density (CIRCULAR FOA-DE-0003033 2024). When E_T approaches 525, performance issues may develop in the battery/cells. This could result in an EV's range degrading to 80%, poor cold weather performance, low power output, and charging difficulties, possibly leading the owner to consider replacement or service. The battery might then be used for other applications. If repurposing costs exceed application benefits, recycling becomes the next option, and its value depends on techniques and cell composition.

Value Chain Feedstock Types

Recycling materials fall into two categories: old and new scrap. Old scrap typically includes batteries used until EOL with $E_T > 525$, while new scrap consists of unused batteries ($E_T \sim 0\text{--}10$), formed cells, printed electrodes, and related manufacturing waste-divertible material.

New Scrap Recycling

New scrap includes materials from manufacturing cells to slightly used batteries with $0 < E_T < 10$. This category covers stages such as dry electrode, slurry, printed electrode, dry cell, wet cell, formed cell, module, battery, pack, and device. The complexity increases at each step and may require specialized recycling. Early-stage materials can be reclaimed near the manufacturing site. Recycling options include pyro/hydro, hydro, or direct methods. Direct recycling is simpler and allows in-house reuse, keeping materials within the manufacturer's control.

Early-stage new scrap includes originally manufactured (OEM) cathodes failing quality checks like calendar life or air exposure, often containing one material – the cathode active material (CAM). This single-component material at the CAM level may have five elements: Li, Co, Ni, Mn, and O.

When mixing the cathode with the binder/solvent solution and conductive additives, two or three components are added. The solvent can be aqueous, organic, toxic, or nontoxic. The mix is printed onto current collectors, dried, and then cut for stacking into pouch cells or rolled into cylindrical cells. Alternatively, dry coating can be used, eliminating the solvent and using a dry printing method. Some methods also eliminate both the binder and the solvent. Wet printing is the most common technique today. Regardless of the method, there will always be some printed scrap material.

Printed cathode foils consist of an aluminum current collector, CAM, a conductive additive, and a binder. The printed scrap material can be recycled with any of the methods. Pyroprocessing uses metallic aluminum present in the current collector as a reduction agent; hydroprocessing removes printed material to dissolve and purify CAM; and direct recycling separates and reuses printed CAM in new mixes.

Printed anode foils consist of the copper current collector, graphite, silicon additive, and binder. Graphite has two general types: synthetic and natural. Silicon has become a common additive to improve capacity. The engineered nature of these parts gives their materials a relatively high value over their elements. Direct recycling offers a possible route for their recovery where hydro may result in irreversible damage to these engineered materials; and pyro would use them as combustible fuel to support metallic reduction of cobalt or nickel oxide components.

The third/fourth stage of new scrap is for dry cells and wet cells. These are cells that have been stacked and made ready for electrolyte addition, but they have a quality assurance failure. Perhaps there is a misalignment of the cathode/anode or a fold in the separator. These cells have not been put through the formation cycle. This level of scrap is very similar in complexity to the printed level; however, there are more components including the separator, printed cathode, printed anode, and package.

Evaluating which recycling method is suitable for unformed cells is an important part of manufacturing design. The potential to reclaim significant value is at stake depending on the methods used. The pyro approach would be able to use the additional carbon graphite as a reduction agent in the smelt which might offset any additional energy requirements; however, with all the added components the concentration of Co and Ni decreased. Hydroprocessing would have to remove the printed material then separate the CAM and graphite, dissolve the CAM, and refine that for salts or PCAM precipitation. Direct would have to separate the printed CAM and graphite (and silicon), and potentially reuse each for a new print mix. For the lithium iron phosphate (LFP) scenario, direct recycling is the only opportunity to conserve the inherent value of the manufactured LFP CAM.

The fifth stage of scrap is for formed cells. These have been placed through the first charge and discharge cycles. The first charge cycle results in irreversible lithium loss from the cathode resulting in a formulation change in the CAM. The anode side has undergone solid electrolyte interphase (SEI) formation on graphite and silicon. With formation, full scrap cells have an $E_T = 0$.

Once a cell is on the E_T scale, after formation, they have the hazardous characteristics of flammability and the potential to be damaged/defective/recalled (DDR). DDR and EOL LIB bring a premium for materials handling and transportation to an outside/destination recycling facility. For $E_T > 0$, any of the methods can be used; however, potential cost savings may be found with one approach over another. For smelting, there is a logistics cost, and the batteries have materials energy that can be used for the smelting charge without added energy. For hydro, there is a logistics cost, and the CAM and anode would have to be separated via leaching or other methods, the CAM would be feedstock for PCAM. Direct methods can eliminate the logistical cost, separate CAM, and anode and potentially resolve the lithium deficiency in the CAM making it suitable for reuse.

Nickel, manganese, cobalt oxide (NMC) has some flexibility for recycling type as there is Ni/Co value potential. The recycling scenario for low-no cobalt/nickel chemistries is different. For example, LFP is a non-fit for smelting, a semi-fit with hydrometallurgy for lithium recovery, and a possible fit for direct methods to recover the whole CAM particle.

Old scrap material has all the manufactured components from cells within a module and packs with unknown state of health, unknown match for manufacturing, and a myriad of other considerations.

Old Scrap Recycling

LIBs have a lifespan of use with $E_T \sim 525$ in their original application, such as EV, and may then transition to a second use application, such as an ESS. The time involved to reach old-scrap EOL could range from 7 to 15 years. Simply speaking, if the scrap value is higher than the repurpose cost, recycle processing is triggered. Some variables impact the repurposed value and cost; the implementation of new technologies will certainly improve the scrap value and the repurposing cost.

At the intersection of recycling and reuse, EOL packs and vehicles may be purchased and sorted for appropriate use or recycling. For high nickel content NMC batteries, the B_{sv} will be related to nickel pricing. Nickel pricing (or battery materials in general) should be low to support an affordable EV market, on the other hand, high nickel pricing supports refining and recycling. The B_{sv} must be positive to move materials toward recycling, whether that value is subsidized, financed, or completely up to the market. For the recycling technologies, each has strengths for economic recovery that can complement one another, so the best management of recycling and decision pathways for processing will yield the highest value. For that scenario to occur, there must be technical options and the ability to access them with operational decision-making. Logistical technologies can assist the transition from use-case to recycle – these include sorting, classification, and safety technologies.

The single commonality for all chemistries is the characteristic hazard of flammability, even for solid-state batteries containing lithium metal. The scrap value $(S_V) < B_{SV}$ is due to logistical costs for managing hazards (L_{SC}). So, $S_V = B_{SV} - L_{SC}$. Therefore, a negative S_V is a problem for an EV; in other words, it is costly to move old scrap batteries that might catch fire. Countermeasures to reduce risk add cost; these include packaging, training, load limits, and insurance: they all add up in the liability column for S_V. It follows that the pretreatment processes described above have a critical opportunity to minimize L_{SC} to maximize S_V. The limit for S_V is B_{SV}; the fundamental materials value in EV and ESS batteries.

Technical Methods

Smelting, hydrometallurgy, and direct recycling methods are introduced in this section. Sidebar discussions of battery BOMs provide a context for what is being recycled, which provides context for the methods.

PRACTICAL INSIGHTS | Cell Chemistry Examples for LIB Bill of Materials

To support the discussion on recycling methods, this sidebar shows a general representative example of a LIB formulated with a cathode based upon NMC. Another example of the cathode material is $LiPO_4$ (LFP) which would contain generally the same other components. A cylindrical cell geometry is shown in Table 5.1, but other examples include prismatic or pouch cells, which are typically made from polyethylene-coated aluminum.

The sample BOM shows a feedstock example for the Recycling Methods descriptions. The methods have different product goals that will be explained in detail in each section. If the intent of materials recovery from recycling is to isolate transition metals from the cathode, that is roughly 59% of the cathode weight or ~17% of the cell weight. The cathode weight contains oxygen which is necessary for the function, but not part of the metallic weight. The elemental lithium content comes from about 7% of the cathode material or ~2% of the BOM. The electrolyte (1 M $LiPF_6$ solution) adds 0.07% to the lithium weight. Lithium recovery is expressed in terms of the recovered

(continued)

> **PRACTICAL INSIGHTS** | Cell Chemistry Examples for LIB Bill of Materials (*continued*)
>
> lithium compound, lithium carbonate, for example. Assuming no loss, 1 g of lithium produces 5.323 g lithium carbonate (for a 5.323 LCE for Li). The LCE is 11 for the BOM. For the recycling methods, pyrometallurgy and hydrometallurgy would have the same maximum value of 11 for the LCE. Direct recycling recovers lithium from the BOM as a cathode compound, so it might not (practically) have an LCE under current assessment methods.
>
> **TABLE 5.1** Bill of Materials Example for Conceptual Discussion
>
> | Component | Material | Percent Weight in the Cell (Rough Example) |
> |---|---|---|
> | Cathode | $LiNi_{0.6}Mn_{0.2}Co_{0.2}O_2$ NMC622 | 30% |
> | Anode | Carbon graphite | 30% |
> | Electrolyte | EC:EMC 1M $LiPF_6$ | 10% |
> | Cathode current collector | Copper (Cu) | 7% |
> | Anode current collector | Aluminum (Al) | 7% |
> | Separator | Polypropylene (PP) | 4% |
> | Cathode binder | Polyvinylidene difluoride (PVDF) | 2% |
> | Anode binder | Styrene-butadiene rubber (SBR) or PVDF | 2% |
> | Cylindrical case | Ni plated steel | 8% |
> | | | 100% |

Pyrometallurgy

Pyrometallurgy, or smelting, recovers valuable metals from LIBs including Co, Ni, Cu, and Li. Li is not the primary recovery goal for smelting but can be economically recycled when the commodity price reaches higher levels. With these limitations, LIB chemistries with low or no cobalt have limited value and can inhibit its recovery. LFP, for example, contains iron, which causes deleterious side reactions that inhibit cobalt recovery in the metallurgy of smelting. It follows that the increased use of LFP and subsequent recycling has restrictions or exclusions from smelting, that is, LFP has limited value and is a liability as smelting feed.

There are two steps in pyrometallurgy: (1) thermal harvesting of harvest transition metals and (2) hydrometallurgy to purify them. In the first step, high temperatures (>1,500°C) transform the metal oxides present in LIBs into metallic alloys. The metallic alloys have a relatively high density and settle into a melt, which can be cast into ingots, cooled, and collected. The alloy is an intermediate product that requires hydrometallurgical refining of Co and Ni as purified metal salts. In contrast to the melt, there is a floating material atop known as slag. Slag contains the light metal oxides of aluminum, lithium, silicon, and calcium. Slag is also an intermediate product that requires hydrometallurgical refining for lithium, which is a separate process from hydrometallurgy for Co/Ni.

Smelters are designed to reach these high temperatures through the combustion of fuel including LIB materials. Combustible LIB materials include scrap with electrolytes, plastic, carbon, and active lithium. In addition to fuel, plasma can be used to increase the thermal energy in the smelting process.

The advantages of smelting are simplicity and environmental controllability of the plant. On the other hand, the need for high levels of cobalt (and nickel) is a challenge with the growth of low and no cobalt LIBs.

Pyrometallurgical Processing Example

The smelting of LIBs in a practical process example contains a "charge" of material consisting of 69% LIBs and 31% fluxing agent composed of 1:1 SiO_2:CaO and aluminum metal from batteries. The Al in batteries is from current collectors and packaging. A fluxing agent is a material that forms and stabilizes the slag layer from the molten metal layer.

The goal of smelting is to eliminate LIB waste and to recover elemental value. The major limitation is recovery of Co and Ni, which complicates the calculus of feeding material into the smelt charge. LFP inhibits the recovery of Co and Ni through the thermodynamics discussed above, complications with PO_4^{3-} chemistry, and nonproductive space in the smelt charge.

The charge is ignited and reaches temperatures from 1,450 to 1,650°C with limited input of energy or materials. If the LIB contains electrolytes, carbon, Al, and other flammables, then they can reach temperature without much external energy input. However, if the charge material contains black mass without the combustible material of a full LIB, then external energy is likely required to reach metallothermic reduction temperatures.

When the charge reaches sufficient temperature the LIB transition metal oxides capable of reduction will form the dense, molten metal (melt-layer). The lighter metals form oxides that partition into the slag layer which floats upon the melt layer. The melt layer is dense and can be drained away from the bath smelter where it solidifies into a metal alloy preliminary product. The alloy is refined into cobalt and nickel salts through hydrometallurgical processing.

The slag layer is removed and used potentially for concrete manufacturing. The elements contained in the slag include oxides and fluorides of Li, Al, Ca, and Si. With a sufficiently high price of lithium, it is economical to isolate it from the slag as carbonate. However, lithium purified from slags requires multiple process steps to extract and purify via precipitation or ion exchange. Lithium purification from hard-rock sources is generally 3–5× more expensive than from brine, so slag processing may have similar economics as the purification of lithium from ores mined from hard-rock sources.

Hydrometallurgy

Hydrometallurgy uses aqueous solutions and organic chemical solutions to refine Co/Ni metals. From the starting point of an alloy or black mass through to the product of refined metal salts and/or PCAMs requires multiple steps: (1) dissolution (2) extraction (3) purification, (4) precipitation, and (5) drying (Figure 5.1).

Hydrometallurgy is widely used in LIB recycling. The engineering draws from the separation of Co/Ni in mining and catalyst recycling. It is employed to refine metallic alloys from smelting or to leach and refine LIB scrap CAMs from black mass. Black mass and alloy refinement are discussed below. There are two general product types from hydrometallurgy: (1) purified metal salts (sulfate or nitrate) and (2) precipitated mixed metal oxides used as PCAM.

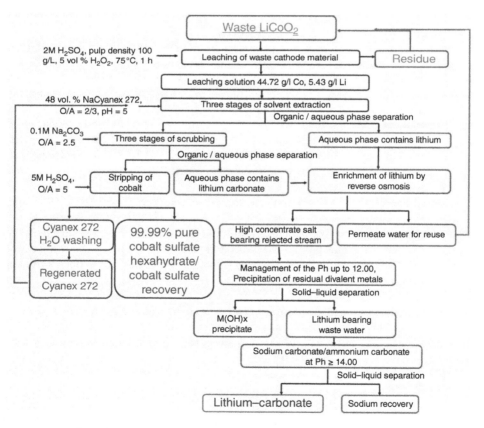

FIGURE 5.1 This schematic lays out the process by which lithium is recovered through hydrometallurgical process. *Source: Swain, B. (2018)/John Wiley & Sons.*

Hydrometallurgy from Alloys

The alloy produced from smelting contains the transition metals from the LIB. Cobalt and nickel are the main elements of interest for separation in the alloy. The separation is accomplished with a liquid/liquid method; therefore, the solid alloy needs to be oxidatively dissolved into a solution. This process involves starting from the metal and oxidizing it to a (2+) ion. Dissolution of metallic species can be accomplished with an oxidizing acid such as nitric, or a mixture of sulfuric and an oxidizing agent. Alloy dissolution is free from organic components, salts, and light elements removed in the smelting process. The organic components interfere with liquid–liquid extraction (LLE), so smelt-to-hydrometallurgy requires a relatively simple process flow sheet.

The LLE occurs through mixing two immiscible liquids: water and kerosene. A complexing agent, such as an alkylphosphinic acid is mixed with the system. Cyanex 272 is an example of a commercial dialkyl phosphinic acid (PA) used for Co/Ni separation. The general formula is $R_2PO(OH)$ with $R=((CH_3)_3C-CH_2-CH(CH_3)-CH_2-)$. The PA remains dissolved in the aqueous phase. When the phosphinate reacts with Ni(II) it produces a polar compound, Ni-PA-H_2O, without affinity for kerosene. On the other hand, when the phosphinate reacts with Co(II) it produces a nonpolar compound, Co-PA, with an affinity for the kerosene phase. The Co-PA partitions into the kerosene and the Ni-PA-H_2O remains in solution thus separating the Co and Ni.

The metal–organic complexes are a long way from becoming a cathode material, they must first be released from their organic complexes. The metal–organic complexes have their phosphinates stripped away via treatment with sulfuric acid, leaving the metal sulfate and PA (which is reused). The metal sulfates are dried and useful for manufacturing. It should be noted that drying metal sulfates is an energy-intensive step that requires brute-force heating of the solution to vaporize bulk water, and more energy to release the hydrated water from metal-salt-hydrate to produce the dry metal-sulfate.

Hydrometallurgical Recycling from LIBs (Without Smelting)

Hydrometallurgical LLE is used with and without smelting. The main difference is the pretreatment of LIBs, without smelting, batteries are shredded or otherwise disassembled into black mass prior to hydrometallurgical processing. Dissolution of the CAM metals from the black mass is accomplished with acid leaching from a heap of black mass or mixing of the black mass with acid. A homogeneous solution of transition metal sulfates and lithium sulfate from CAM is accomplished by heating (~80°C) a solution of sulfuric acid and hydrogen peroxide. The hydrogen peroxide is a reducing agent for the transition metal (3+) to produce a soluble (2+) ion.

PRACTICAL INSIGHTS | Hydrogen Peroxide Decomposition

Hydrogen peroxide has oxygen in the (−1) oxidation state, usually oxygen is (0) for O_2 or (−2) as in H_2O. Hydrogen peroxide decomposes into water and oxygen, which means it can act as an oxidizing agent by accepting electrons from another element/compound or it can act as a reducing agent by donating electrons to another element/compound. It is why hydrogen peroxide/sulfuric acid dissolves CAM by reducing the Co^{3+} (for example) to Co^{2+}; it is also why the acid/hydrogen peroxide dissolves metals by oxidizing them to the (2+) state.

Hydrogen peroxide is an important chemical used in LIB recycling for classic hydrometallurgical processing to dissolve metals. It is also used in direct recycling to stabilize (3+) metals in the CAM and assist in the dissolution of trace metals.

Electrochemical Hydrometallurgy

Dissolution of CAM can be a chemically intensive way to perform reduction; however, electrochemical methods can be used to achieve dissolution and there is a robust body of work developing the approach. Such developments feature minimization of the use of chemicals replaced by the use of applied voltage in a reaction cell. The benefits can be (1) savings in the cost of chemicals and their environmental management and (2) flexibility for power sources to operate the electrochemical reaction including (a) off-peak power or (b) carbon-free sources.

Once the CAM transition metals have been dissolved, the options for the next steps are to (1) purify and refine the individual metal salts using LLE or (2) purify the mixture of metal salts and reprecipitate the CAM, which is discussed in the next section.

Reprecipitation of Cathodes from Hydrometallurgical Processing

This section discusses making new LIB cathodes by using the dissolved solution from old cathodes or formulating the purified metal sulfate in new manufacturing. To prepare the reader for the concept of reprecipitation of cathodes, the following sidebar discusses a general synthetic consideration for lithium-mixed metal oxide cathodes. This is to provide context and understanding LIB recycling strategies.

PRACTICAL INSIGHTS | Coprecipitation for Cathode Synthesis

Commercial manufacture of mixed metal cathodes, such as NCA, containing nickel, cobalt, and aluminum has been practiced since the late 1990s and NMC since the 2000s. Generally, these are made through coprecipitation of the transition metal oxyhydroxides from a solution of the $M(II)SO_4$ solutions (M=Co, Ni, Mn, and others). There are many coprecipitation strategies and techniques developed to manufacture cathodes with structural properties that promote features such as thermal stability. Notably, for the synthesis of NCA, precursors of Ni(II) and Co(II) sulfate are coprecipitated prior to the addition of aluminum to suppress the propensity of Ni to form cubic NiO. In any case, the product of coprecipitation is a mixed oxy-metal hydroxide that is commonly referred to as a PCAM. Classic PCAM precipitation produces sodium sulfate waste products. Sulfate waste recycling is a cost barrier for implementing cathode manufacturing. Disposal of sodium sulfate solution into large water bodies, such as oceans, is the lowest cost option and is generally allowed in locations with little regard for best practices or environmental health and safety regulations.

The PCAM is transformed into a CAM by solid-state reaction with a lithium source such as lithium carbonate or hydroxide. The solid-state reactions can require many hours of heating at high temperatures under cover of oxygen (for NMCs).

Removal of sulfate in the coprecipitation equation has been achieved by using metal nitrates as the precursor materials. Direct synthesis is practiced by feeding lithium nitrate and mixed-metal nitrates into a plasma furnace where they form the CAM. Plasma and reaction control can be tuned to produce single-crystal CAM or polycrystalline CAM.

The metal sulfates or metal nitrates can obviously be used to prepare PCAM and CAM. The source of these salts can be from recycled battery materials. From the perspective of the recycling processing industry, these are product entry pathways. Recycled material can be metal sulfate or nitrate; mixed metal sulfate or nitrate; PCAM; and CAM.

The opportunity occurs to produce NMCs from the M(II) sulfate solutions made in the hydrometallurgical recycling of LIBs. After dissolving the harvested batteries, the solution requires purification of Al, Cu, Fe, Li, and other metals to reach an acceptable NMC precursor solution with Ni(II), Mn(II), and Co(II) sulfate in the ratios required. The approaches to purify solutions use pH adjustments to precipitate and filter out impurities. After achieving that precursor purity and concentration, the NMC synthesis follows known manufacturing procedures.

A feature of this approach is the flexibility to recycle LIBs and achieve a controlled mixture of metals in solutions. The benefit of reprecipitating cathodes from a mixture produced from recycling is to have a useful product that avoids the cost of extensive purification to the elemental level (i.e. purified metal sulfate salt). As these materials are used in NMC cathode manufacturing anyway, the approach avoids the energy used for purification of the $M(II)SO_4$ followed by remixing into much the same type of metal solution for NMC synthesis. It helps to avoid industrial waste from unnecessary redundancies.

Pretreatment and Harvesting

Dismantling exposes the battery active electrode material (BAM) or "black mass" from the manufactured components. There are many ways to transform batteries into black mass and they involve shredding. Shredding batteries will cause thermal events and battery fires. The approaches to mitigate fire include the following shredding strategies:

Cryo-shredding

Pretreatment of cells and packs by soaking in liquid nitrogen makes the components cold enough to slow the reaction during physical processing. However, as the components warm, their reactivity accelerates. This is commercially applied to lithium primary batteries, where the frozen cells are shredded into an aqueous solution designed to neutralize thionyl chloride or sulfur dioxide electrolytes. The approach can be modified for LIBs with organic electrolytes based on carbonate solvents and $LiPF_6$.

Aqueous Shredding

Shredding LIBs at room temperature under and into an aqueous solution provides a heat sink for the thermal events. The solution chemistry contains basic materials to neutralize acids (HF, $HOPF_x$'s) produced from the decomposition of LIB electrolyte salt, $LiPF_6$. Even with the mitigating approaches, the shredding action and resulting thermal events produce temperatures >1,000°C immediately at the cutting site before reaching equilibrium with the aqueous bath.

Discharging batteries prior to shredding may reduce the severity of thermal events; however, discharged (or deactivated) LIBs remain with flammable characteristics even under the best circumstances. Discharging batteries takes time and measurements that may be considered too costly. Due to uncertainty of the state of charge (SOC), shredding systems are being designed to process charged materials; however, these shredders are exposed to thermal runaway events (albeit localized and concentrated within the shredder). Such events can lead to localized temperatures >1,000°C, which can degrade shredder components. The longevity of what should be durable capital equipment under such stress remains to be fully understood.

Dry Shredding

LIBs that have been discharged to short circuits may be shredded with a cover gas of nitrogen or carbon dioxide (or air), or under vacuum. The probability of incidents remains finite; however, since LIBs (1) may not be able to be discharged, (2) will rebound from a discharged state, and (3) may be charged but show a discharged state due to circuitry malfunctions within the cell or on the battery.

Removal of electrolyte solvents can be performed before shredding on punctured batteries or afterward on the shredder residue. Electrolyte extraction can be accomplished with common physical chemical methods including vacuum drying or solvent extraction. Patented systems have been developed for electrolyte extraction by employing supercritical carbon dioxide (for example, Sloop 2003, Grützke et al. 2014). Vacuum drying removes volatile and flammable electrolyte solvents. Vacuum drying is well established industrially for the removal of volatile organic components (Burch 1929) and has found application in many industries that use volatile organics, including with LIB recycling for the removal of volatiles from batteries and shredder residues.

Shredder residue from LIBs may still harbor risks with moderate heating. Workers at Münster Electrochemical Energy Technology (MEET) recreated situations in which shredder residue may be sealed and heated due to environmental conditions. They observed fluorophosphate decomposition products at moderately high temperatures (Grützke et al. 2015). Obviously, the removal of organic components prior to heating avoids their subsequent reactivity (including fire) to produce toxic material. The toxicity risk, beyond the flammability risk focused on here, underscores the need for controlled extraction with a CO_2 system.

Hammer Milling

Shredding of batteries produces parts related to the size of the blades. A first pass on LIBs may produce 1–2 cm^2 pieces of laminated and encapsulated materials. Further size reduction and liberation of electrodes encapsulated within folds made through the primary shredding step. Granulators, or hammer mills are used to pulverize the LIB materials into small particles.

Thermal Treatment

Heating LIB material to the range of 300–600°C will burn away organic material including (1) electrolyte, (2) separator, (3) active lithium within graphitic carbon or silicon, (4) polymer-coated aluminum and other packaging materials, and (5) polymeric binders – the glue holding the electrode material to the current collector. Reaction 1 shows the heating decomposes the PVDF binder to leave the cathode, anode, and metallic components.

Reaction 1:

$$NMC : PVDF : Al \xrightarrow{heat} NMC + Al + H_2O + CO_2$$

HF is produced from the decomposition of the binder and from the electrolyte salt LiPF$_6$. HF will fluorinate the cathode as a side reaction, neutralization of the acid can be achieved with a suitable base.

Thermal treatments leave a relatively clean electrode surface that is free of binder, nonvolatile electrolytes, salts, and the solid electrolyte interface. In theory, it simplifies the mixture into two components, the cathode, a lithium metal oxide, and the graphitic anode. The anode could also be a silicon compound. In practice, the HF fluorinates metals, the fluoride becomes difficult to remove in subsequent processing. However, thermal treatment does expose the surfaces of each electrode material: CAM and anode. This enables floatation through the exposure to the hydrophilic character of CAMs and hydrophobic character of graphite anodes.

If the next step is to separate the cathode from the anode with froth floatation, clean particle surfaces are critical for adsorption of the floatation agents. Aliphatic compounds will have good adsorption to a graphitic surface and increase the hydrophobicity and ability to float in water.

If the step after the thermal treatment is to dissolve the lithium metal oxide, there is no barrier keeping acidic contact with the oxide.

Froth Floatation – Separation of Carbon from Metal Oxides

Froth floatation is a physical method to separate materials based on their difference in hydrophobicity. Hydrophobicity is a surface characteristic property. Regarding particles, it is impacted by the type of elements and chemical bonds present at the surface. For LIB cathodes, for instance, the surface is composed of metal–oxygen bonds which have a polar character; these will have attraction for water and, therefore are hydrophilic. Graphite, on the other hand, has carbon–carbon bonds present on the surface, these nonpolar bonds have no attraction for water, and they are hydrophobic. Given the density of graphite is 2.2 g/cc and water is 1 g/cc, it seems that the material would sink in water; however, the hydrophobicity of the surface can make it float. The hydrophobicity can be accentuated with frothing, so mixing air into a mixture of graphite/water allows bubbles to adhere to the graphite surface and float to the surface. Froth floatation is useful for separating metal oxides from carbon–graphite in LIBs where carbon can be made to float (Vanderbergen et al. 2022).

As with any physical method, separation is incomplete. Therefore, small amounts (%) of graphite remain mixed, and vice versa with metal oxide in the graphite float. For LIB recycling pretreatments can be developed to accentuate the surface characteristic difference between cathode and anode to allow graphite floatation. The separations technique is challenged in situations with cathodes that have carbon coating (for example, LFP) and for anodes made from metal oxide such as lithium titanium oxides (LTO). Anodes with silicon add another complexity, but the high density of silicon compounds allows them to sink in a density method. For the recovery of graphite and highly engineered silica compounds in anodes, floatation may provide a viable economic pathway for collection of the material.

Reprecipitation of Cathodes from Hydrometallurgical Processing

This section discusses making a new CAM by using the hydrometallurgical dissolved solution from old cathodes. This strategy avoids the isolation of elemental salts through LLE steps. Instead, the metal solution can be purified to remove unwanted trace elements and adjusted with the addition of metal sulfate to prepare it for manufacturing of a PCAM. In this strategy, hydrometallurgy is used to prepare a solution for manufacturing.

For example, NMC can be made from the M(II) sulfate solutions produced from the dissolving of LIB scrap or black mass. After dissolving the harvested batteries, the solution requires purification of Al, Cu, Fe, Li, and other metals to reach an acceptable NMC precursor solution with Ni(II), Mn(II), and Co(II) sulfate in the ratios required. The approaches to purify solutions use pH adjustments to precipitate and filter out impurities. After achieving that precursor purity and concentration, the NMC synthesis follows using known manufacturing procedures for adjusting pH to precipitate NMC as a PCAM. The PCAM is transformed to CAM through calcination with a lithium salt and oxygen cover gas.

A feature of this approach is the flexibility to recycle LIBs and achieve a controlled mixture of metals in solutions. The benefit of reprecipitating cathodes from a mixture produced from recycling is to have a useful product that avoids the cost of extensive purification to the elemental level (i.e. purified metal sulfate salt). As these materials are used in NMC cathode manufacturing anyway, the approach avoids the energy used for purification of the $M(II)SO_4$ followed by remixing into much the same type of metal solution for NMC synthesis. It helps to avoid industrial waste from unnecessary redundancies.

Direct Recycling

The direct recycling approach is to minimize energy and material input (cost) in recycling. Direct recycling maintains the electrode particle without decomposition and separation into constituent elements and purification of trace metals. Decomposition of CAM crystal lattices in recycling relates to inputs consistent with high-temperature treatment, or chemically intensive hydrometallurgy, for which the cost scales with chemical and energetic intensity. Preservation of the crystal lattice preserves the material and energy inputs to originally produce that crystal lattice. The energy and material inputs can be limited to repair the CAM lattices, direct recycling energy inputs (for CAM repair) can be a fraction of the energy and material input for original syntheses.

The direct approach conserves and repairs the structural characteristics of electrode particles, especially the lattice, the original formation of which is energy intensive. Cathode healing™ stands out among the direct approaches to conserve the lattice and clean it – that is, remove trace metals, and achieve the high purity and electrochemical performance required for battery manufacturing.

Synthetic strategies used to produce lithium transition-metal oxide electrode materials in the first place can be applied to repair them. A nonexhaustive list includes classic solid-state, electrochemical, hydrothermal, nonaqueous molten salt, sol–gel, and ion-exchange synthetic methods. All can be applied as direct recycling approaches to reset or make a new cathode material. These solid-state, soft-chemical synthetic methods offer fruitful techniques and strategies for recycling and have been adapted and applied to the field of Direct Recycling (Sloop 2009b, 2016a, b, 2017a, b, c, 2018a, b, 2019a, b, c, d; Sloop and Allen 2014; Sloop et al. 2020).

Solid-State Synthesis

As a direct recycling approach, solid-state methods top-off, reintroduce, or replace the lithium that goes missing from the active complement in the original manufacture of the cathode. This requires a stoichiometric addition of lithium using a salt, relative to the amount of lithium "missing" from the spent lattice. Also, that process "resets" transition metal oxidation numbers to "original-manufacturing", for example, Co(III) in lithium cobalt oxide (LCO), and Mn(IV,III) in spinel.

Solid-state heating has limitations for direct recycling including unintended reactivity of spent cathode, repair of a CAM with an irreversible failure mechanism, and mixtures of different cathode materials all can yield products other than a functional electrode material. A notable issue is a common item found with recycle-harvested electrodes, the polyvinyl difluoride polymer that binds cathode-carbon mixtures. When heated, it produces hydrogen fluoride that attacks the cathode. Moreover, common are cathode failure mechanisms linked to transition metal dissolution. When transition metals go missing from the lattice, they cannot easily be reinstated.

Electrochemical Reintroduction of Lithium to Spent Material

Even the simple, obvious task of discharging a battery is an act of direct recycling. The electrode particles inside the cell take in lithium while reducing the oxidation state of the transition metal oxide host. However, this has limitations due to the battery management system preempting full discharge, and how the lithium goes missing from the lattice in the first place. Such irreversible loss of lithium inventory occurs during the very first charge–discharge cycle in the life of the battery, and during the life of the battery with inevitable side reactions that trap lithium in the lattices, precipitate them from solutions, or paste them on the surface of electrodes as "solid-electrolyte interfaces." For complete direct recycling, the reduction of the CAM requires an input of lithium.

For spent electrode materials harvested from EOL cells, their electrochemical reduction is achieved through a cathodic reaction in an external, electrochemical reactor. The system has a working electrode that contacts the spent electrode material, electrolyte (non-aqueous, aqueous, or even a molten salt), and counter electrode to provide lithium ions. Electrochemical reintroduction of lithium is akin to the discharge reaction in a lithium-cell except with an excess supply of lithium. While electrochemical reintroduction of lithium can address some lithium inventory deficiencies in spent electrodes, it may not be able to address cation mixing of transition metal ions with lithium ions, or transition metal dissolution from electrodes.

Hydrothermal/Ionic Solution Methods

Hydrothermal methods have been developed to recycle CAM electrode particles. In a typical example, the spent CAM electrode is soaked in a lithium-containing solution using any suitable solvent (aqueous/organic), ionic liquid, or molten salt. The mass action of lithium upon the CAM results in the reestablishment of lithium inventory in the CAM. After the ionic treatment, a final regeneration of the CAM may be achieved with calcination.

An example of hydrothermal treatment is for CAMs based on LCO. Spent LCO may have lithium deficiency of 10–15%, and structural changes from layered to cubic nature. Such structural changes impede the transport of lithium to and from the CAM host.

There are four advantages of direct recycling: (1) It avoids multiple dissolution/precipitation steps to make a PCAM and the final high-temperature treatment CAM. It requires as little as 10% of the energy required for NMC CAM from purified inorganic salts. (2) It reuses the lithium in the cathode. (3) It enhances the flexibility to chemistry by reinstating the functionality of an electrode, rather than purification of elements in the electrode. (4) Direct methods are applicable to cobalt/nickel-free chemistries such as LFP.

Battery Pretreatments

Battery pretreatment is aimed at improving the safety of EOL handling. There are three general approaches practiced in the industry. (1) Block the electrodes so that a short circuit is avoided. (2) Package cells and batteries sufficiently so that in case of a thermal event, the packaging will absorb heat and stop propagation of any fire. (3) Remove the residual charge from batteries to reduce the potential energy released in case of a short circuit.

The first two approaches are related in that the battery can remain in a charged state. Physical blocking of the electrodes can be achieved by the application of tape to the leads. This is labor intensive but does not require skill to succeed. Specialized containers for batteries can be designed to hold individual cells/modules where they cannot move into a short circuit. They are easy to load and provide thermal and physical shields against significant collision or fall.

Discharging batteries controllably removes the stored electrical energy. Any amount of stored energy may be released rapidly to create sparks and heat that can lead to fire. Discharging batteries is achieved by connecting the positive and negative electrodes through a circuit that performs work or produces heat. Such a circuit could be a solid-state resistor, a positive thermal coefficient resistor, ion-containing solution, and/or a short. While it seems simple to discharge LIBs there are complications in achieving discharge, these include (1) battery management systems that prevent the full discharge of cells (2) unbalanced cells in a battery, and (3) cell-internal faults (i.e. low state of health). With the sheer number of batteries and cells, considering the potential complications, there is certainty that charged cells will exist after discharge deactivation at EOL. A discharged LIB still contains flammable material which qualifies it as a Class-9 hazard.

EOL LIBs are miscellaneous (Class-9) hazardous materials. They are managed as Universal Waste since they are ubiquitous manufactured goods with measurable, limited, controllable hazardous characteristics. An example of a characteristic hazard within an EOL LIB is related to flammability as the electrolytes have a flashpoint <140°F. Toxicity and corrosivity characteristics can be changed with neutralization; this give LIBs changeable characteristics unlike listed hazardous characteristics from toxic metals such as lead (Pb) or mercury (Hg).

There are approaches to eliminate characteristic hazards. For example, battery neutralization is a patented process (Sloop and Crandon 2020) that removes flammable characteristics from EOL LIBs. As the EV and ESS industries grow, successful neutralization will eliminate the risk of battery fires and associated costs that inhibit the successful growth of the EV and ESS. The current risk profile for EOL LIB is

FIGURE 5.2 Fire triangle components of oxygen, heat, and fuel. LIBs contain all three manufactured together. *Source: Meduza / Adobe Stock Photos.*

$36,000/ton of spent batteries (Callaway et al. 2022), which is related to fires, injury, downtime, and property damage.

Understanding the fire triangle is essential when assessing the risks associated with LIBs, which inherently contain the three components required for combustion: ignition, fuel, and oxygen (Figure 5.2). Fuel is from the electrolyte and the lithium; heat for ignition is from a spark that could occur between the cathode–anode; and finally, oxygen is released from the cathode with heat, and it is available in the air should the LIB become breached.

As part of an effort to develop the U.S. supply chain in recycling critical battery materials and manufacturing batteries, the U.S. Department of Energy has prioritized ways to *"improve the safety of EOL EV battery transportation by preprocessing the battery to allow the resulting product to be transported with less restrictive designations at lower cost"*. The use of CO_2 as an inertization fluid addresses *"designs that contain a cost-effective thermal runaway protection or fire suppressant system."* Application of inertization is a potential *"logistics solution that combines storage with transportation to allow full truck load (FTL) shipping"* (Bipartisan Infrastructure Law [BIL] FY23 BIL n.d.).

Spoke Deactivation

Successful transformation of the hazardous classification of EOL LIB can have a major impact on the cost and safety of recycling. The cost driver is compliance with storage and transportation requirements for Class-9 hazardous material, which includes special packaging and procedures to address potential EOL battery thermal runaway. It follows that the elimination of the characteristic hazards reduces their management burden. Successfully removing Class-9 hazardous characteristics changes logistical cost and risk for transportation and storage of EOL LIBs.

The global lithium-ion battery recycling market is valued in the billions of dollars and is estimated to grow at a CAGR of 20.6% from 2024 to 2032 (Global Market Insights October 2024). Lithium-ion cost is $120/kWh, and >40% of the value is cathode, which is a highly engineered material containing critical elements such as cobalt, nickel, and lithium. EOL batteries will become a commodity source in the domestic supply chain for lithium-ion battery manufacturing (LiBridge 2025).

Recyclers experience lithium-ion **fires** (Verzoni 2024) during storage, handling, and shipment. Battery fires cause injury, downtime, and capital equipment loss that is a challenge for insurance coverage.

Successful neutralization of EOL LIBs will transform shipping costs from $2/kg to $0.50/kg and reduce the cost burden to less than 10% of the total cost of recycling, which is critical for the affordability of EVs and large format batteries.

Sloop and Crandon patented a method to render batteries inert with the treatment of carbon dioxide. Validation of the nonflammable and nontoxic characteristics recently led to the interpretation from the Pipeline Hazardous Materials and Safety Administration (PHMSA) that LIBs that have been neutralized through the patented process are not considered hazardous materials. The process is designed to neutralize batteries at or near the place they are deemed EOL, thus eliminating the consolidation and shipment of hazardous materials.

In 2024, the State of Illinois Battery developed a model for recycling LIBs, which was in response to the need for a redesign of management after improper handling/storage led to fires (Cassel 2024). The Act:

- Creates an expanded producer responsibility (EPR) program for small and medium format batteries
- Establishes consumer battery recycling goals (rechargeable batteries 60%, all other primary batteries 70%)
- Establishes collection network standards (1 permanent collection site within a 15-mile radius for 95% of the state's population)
- Establishes at least 1 collection site or event for every 30,000 residents in a single county

The application of battery neutralization on a scale of 1 neutralization site: 30,000 residents is an example of how neutralization can be applied as spokes to eliminate hazardous batteries before they are consolidated for recycling.

Design and Recycling

The rich variety of features in LIBs has been developed to meet the demands for safety, cost, and performance in the large format LIB market. Manufacturing LIBs for a fleet of EVs establishes a lifecycle risk and enables the practicality of technical design for new-and-old scrap recycling. For an environmentally and economically sustainable LIB market, the cost of recycled LIB material must be less than the originally mined materials. This is necessary to be competitive in manufacturing new LIBs and to achieve a mass balance for reutilization of old scrap materials.

Design features for recycling can be assessed by resource recovery efficiency with respect to recycling technologies. The design variable list is long and complex including cathode, anode, binder, mixed chemistry, electrolyte, additive, cell geometry, packaging, state of health, SOC, and things unique to each manufacturer. The list of possible variables improves the potential for design success with benefits being improved safety and cost reduction.

Materials

Cathode Electrode

The first major commercial lithium battery CAM was LCO, which is composed of 50% Co, 7% Li, and 43% oxygen. It has a high value due to the presence and maximum concentration of Co. The LIB-positive electrode is composed of CAM, conductive additive, electrolyte, polymeric binder, and current collector. The typical binder is polyvinyl difluoride. It is mixed with the parts to make an electrode that adheres to the aluminum foil current collector.

EV batteries from gen-1 Tesla's used LCO and pyrometallurgical recycling at the time were able to access some value from the cobalt recovery. This allowed the industry to establish acceptable minimum limits for cobalt feed into smelters. The high cost of Co is a challenge for producing low-cost vehicles. Low and no cobalt CAMs are used for EV mass production, which may approach cobalt content limits for smelting. Low-cost EV cathode electrodes have low-value potential with existing recycling technologies, which may yield an end-of-life liability (EOLL).

The potential for high-value recovery and end-of-life asset (EOLA) is enabled with the application of direct recycling. This is because the CAM recovered from direct recycling has greater value than CAM produced from elements recovered in other methods. The binder choice can reduce the EOLA from any method and enables EOLA with direct recycling. If a design can yield low-cost, high-performance, safety, and recyclability; binder choice is an example.

Electrode designs that can lead to EOLA include dissolvable ionic crosslinked polymers (DICP), soluble binder, and binder-free slurry electrode batteries. These are summarized below.

DICP

The fabrication of high-energy LIBs depends on the incorporation of the energy materials – namely, the cobalt oxide cathode and graphite anode materials – into battery electrode sheets, which constitute the actual energy storage components inside the battery. Because the energy materials are free powders, a polymer glue (or binder) is needed to process the powders into thin electrode sheets to create the battery. However, having the energy materials glued together by the binder poses significant obstacles for battery recycling, hindering the retrieval of the high volumes of cobalt oxide cathode and graphite anode from spent batteries needed to sustain the growing battery industry.

A "reversible" binder glue that can quickly release the different components of battery electrodes into their individual forms for low-cost, convenient recycling could significantly facilitate recycling of battery materials. An example of such a binder is produced with a two-component polymer system – polyacrylic acid (PAA) and polyethyleneimine (PEI) – whose chemical crosslinks can reverse and disconnect under conditions that are not possible from previous binders. It can dissociate under alkaline conditions to quickly release the energy materials upon recycling, and to readily cure upon drying when fabricating new batteries.

Notably, the binder is free of fluoro-components (PFAS free), so makes it a much greener process than that of Polyvinylidene fluoride (or polyvinylidene difluoride) (PVDF)-containing binders. With many countries contemplating the elimination of PFAS molecules in manufacturing, developing alternatives in battery manufacturing is important. The development is not simple in that PFAS molecules have performance characteristics that align with high voltage, electrochemical, and thermal stability.

With the reversible binder, a facile strategy is employed to recycle CAM and anode electrode materials from spent batteries by utilizing a DICP binder. The three-dimensional network structure of DICP is formed by crosslinking the PAA and PEI through a carboxy-amino ionic interaction. This interaction is pH controlled, and therefore, the crosslinking is readily dissociated in basic conditions, which enables a plausible approach to recover CAM through washing with water.

Soluble Binders

The classic binder used in LIBs is PVDF (as noted above) and to print the electrode, a typical solvent is n-methyl pyrrolidone (NMP). Due to regulatory changes for the elimination of PFAS, the use of PVDF may be phased out in future LIB manufacturing. Also, due to cost, environmental health, and safety concerns, NMP use is not considered to be an industrial best practice. Recycling is an approach to mitigate the potential environmental concerns regarding PVDF. The industry is producing modified PVDF to enable recycling and manufacturing (Amin-Sanayei and He 2015). Modified PVDF can negate the need for organic solvents such as n-methyl pyrrolidone in original manufacturing. It also allows for the use of aqueous methods in recycling. The water-soluble binder has three recycling features: (1) decreased carbon footprint by up to 1/3 by eliminating the need for thermal treatments or organic solvents; (2) PFAS recovery and recycling through simple removal and subsequent processability, and (3) improved safety by avoiding generation of hydrogen fluoride in thermal processing for recycling.

Binder Free, Slurry Electrodes

Battery electrode architecture can be achieved without a polymeric binder. In this case, the cathode is a slurry of electrolyte, CAM, and conductive carbon black. The geometrical format is commercially limited to pouch cells, and several OEMs have licensed the technology from the developer (Woyke 2016).

Binder-free batteries and batteries with releasable binder have very similar characteristics in recycling. These similarities are: (1) the CAM is easily released from the cathode with processing in aqueous media; (2) the potential for binder-fluoride contamination is eliminated; (3) metal current collector contamination is greatly reduced. The foil-electrode adhesion is fluidic in a slurry electrode so removing the electrode does not take pieces of foil with it as does occur with polymeric (glue) adhesion between electrode and foil.

The design features for recycling batteries and electrodes utilizing a releasable binder such as DICP or binder-free electrode slurry include:

(1) High-purity harvested materials:

Fluoride contamination in black mass can be great enough to disqualify hydrometallurgical refining. The fluoride comes from either the electrolyte salt or the PVDF binder. Electrolyte salt is easily removed and can be managed in either pretreatment or refining. PVDF removal by thermal methods contaminates black mass with fluoride. PVDF is difficult to remove in pretreatment and refining, so the added capital cost of process operational units and their operational cost pushes the problem of fluoride contamination acceptability to the stage of PCAM manufacturing. Eliminating the binder or using a water-soluble binder in the cell design essentially removes the fluoride management problem in recycling.

(2) Regulatory compliance and/or Environmental Social Governance (ESG) compliance. Elimination of PFAS – polyfluorinated alkyl substances – in manufacturing.

PFAS materials used in battery manufacturing include PVDF binder, fluorinated molecules (such as fluoroethylene), and some electrolyte salts such as LiTFSI. The support for the elimination of PFAS is broad based on potential regulation, investor expectations, and industrial best practices. There is a movement to eliminate PFAS in manufacturing. If and when that happens, batteries will need to be made without PVDF. So, there must be a viable alternative for electrode design. In this case, there is an overlap between design for recycling and manufacturing requirements.

Battery Service
Batteries are made for single use for as long as possible, up to a million miles are claimed for some LIB formulations. It shows the durability of these materials in a manufactured cell. However, the concept of providing service to cell during life, to rejuvenate the battery, has gained interest from major manufacturers (Agatie 2024) and technology development efforts in the US (ARPA-E CIRCULAR 2024) and China (Chen et al. 2025).

Concept
Battery life limitations are related to physical changes in cathode structure, lithium precipitation from the cathode/anode shuttle, anode physical changes, and electrolyte decomposition. Many of these failure modes have been optimized through years of research and development. Of all modes, LIB electrolyte decomposition contributes to lifetime degradation and may be the limiting factor (Sloop et al. 2003). It follows that removal and replacement of the electrolyte can be a way to extend battery life; this is somewhat like with an internal combustion engine (ICE), the service of an oil change can extend the life of the vehicle. Toward this rejuvenation concept, Sloop teaches methods to remove electrolyte and decomposition products using liquid or supercritical CO_2 followed by replacement of the electrolyte and active lithium ions (Sloop 2003, 2004, 2007, 2009a). The method has been shown to revive LIBs to 90% of their original capacity and continue with a similar fade rate. The implication is that such a service could more than double the life of a common LIB. Within the business model for second use, service, and leasing, perhaps rejuvenation will assist in minimizing the cost of energy delivered over the lifetime of a LIB.

Future Considerations

The environmental and economic sustainability of EVs and ESS requires innovative lifecycle technologies that address safety and cost. Battery neutralization, direct recycling, and battery rejuvenation offer the industry a way to reach affordability of cell lifetime energy cost, LC/ $/kWh <$0.008, which is transformational in making EV and ESS affordable for everyone.

Getting to this ecosystem won't be simple, however. Here are some considerations:

- Transparency as to the origin and manufacturing of battery materials will be necessary. This means instituting a consistent, verifiable means of reporting and, potentially, an auditable, "master repository" of battery data;
- The European Union (EU) and China both have regulatory requirements associated with the disclosure of battery materials while the United States has yet to enforce any such requirements and relies instead on industry standards;
- A recycling ecosystem will require a workforce properly trained and even certified to perform operations with volatile materials safely and efficiently.

Words to Know

Black mass battery-active electrode material (BAM) containing anode and cathode materials.
CAM cathode-active material.
Direct recycling recovers battery materials without breaking down chemical structure.
Fluxing agent material that forms and stabilizes the slag layer from the molten metal layer.
Froth floatation a physical method to separate materials based on their difference in hydrophobicity.
Hydrophobicity the property of being repelled by water.
Hydrometallurgy a method of extracting metals using liquids.
New scrap unused batteries and battery materials.
Old scrap includes battery materials used until their end of life.
PCAM precursor active material, a powder nickel, cobalt, or other chemical elements.
Slag nonmetallic byproduct of metallurgical processes such as smelting.
Stoichiometric quantities of reactants in ratios, as prescribed by an equation or formula.

Further Reading

As with other topics in this book, the scientific literature is dynamic. Argonne National Laboratory makes available an excellent collection of materials, ranging from fact sheets to research reports, related to battery recycling, at **https://www.anl.gov/esia/reference/lithiumion-battery-recycling-publications**

NAAT Baat International also aggregated research on battery lifecycle management at **https://naatbatt.org/research/**

References

Agatie C. Toyota Finds Solution To Rejuvenate Used Li-Ion Batteries, It's Game Over for ICE Vehicles, Published: 7 Jun 2024, 21:18 UTC.

Amin-Sanayei, R. and He, W. (2015). Application of polyvinylidene fluoride binders in lithium-ion battery. In: *Advanced Fluoride-Based Materials for Energy Conversion*, 225–235. Elsevier.

ARPA-E FOA-DE-0003033 (2024). *Catalyzing Innovative Research For Circular Use Of Long-Lived Advanced Rechargeables (Circular)*, 31 January 2024, p. 20.

Bipartisan Infrastructure Law (BIL) FY23 BIL (n.d.). Electric Drive Vehicle Battery Recycling and Second Life Applications. DE-FOA 0003120.

Burch, C.R. (1929). Some experiments on vacuum distillation. *Proceedings of the Royal Society of London Series A, Containing Papers of a Mathematical and Physical Character 123* (791): 271–284.

References

Callaway, J., Navarrete, J., Kasulaitis, I. et al. (July 2022). Lithium Battery Deactivation for Defense Logistics Agency. Energy Academic Group, Naval Postgraduate School.

Chen, S., Wu, G., Jiang, H. et al. (2025). External Li supply reshapes Li deficiency and lifetime limit of batteries. *Nature 638*: 676–683.

Cassel, S. (12 August 2024) **https://productstewardship.us/press_releases/illinois-enacts-nations-16th-battery-recycling-law/**

Department of Energy (DOE) Office of Energy Efficiency and Renewable Energy (EERE) (2 May 2022). Bipartisan Infrastructure Law (BIL) Electric Drive Vehicle Battery Recycling and Second Life Applications. DE-FOA 0002680.

Global Market Insights (October 2024). Lithium-Ion Battery Recycling Market Size - By Chemistry (Lithium Nickel Manganese Cobalt Oxide (NMC), Lithium Iron Phosphate (LFP), Lithium Cobalt Oxide (LCO)), By Process, By Source, Regional Outlook & Global Forecast, 2024–2032. **https://www.gminsights.com/industry-analysis/lithium-ion-battery-recycling-market**

Grützke, M., Kraft, V., Weber, W. et al. (2014). Supercritical carbon dioxide extraction of lithium-ion battery electrolytes. *The Journal of Supercritical Fluids 94*: 216–222.

Grützke, M., Krüger, S., Kraft, V. et al. (2015). Investigation of the storage behavior of shredded lithium-ion batteries from electric vehicles for recycling purposes. *ChemSusChem* 8: 3433–3438.

Langley, J. (1 July 2021). **https://www.letsrecycle.com/news/uk-narrowly-misses-2020-battery-target/**

LiBridge (2025). Bridging the U.S. Lithium Battery Supply Chain Gap. **https://www.anl.gov/li-bridge/**

Sloop, S.E. (9 January 2003). System and method for removing an electrolyte from and energystorage/or conversion device using a supercritical fluid. US Patent 7,198,865.

Sloop, S.E. (4 December 2004). System and method for processing an end-of-life or reduced performance energy storage and/or conversion device using a supercritical fluid. US Patent 8,067,107.

Sloop, S.E. (7 March 2007). System and method for removing an electrolyte from an energy storage and/or conversion device using a supercritical fluid (7 March 2007). US Patent 7,858,216.

Sloop, S.E. (28 July 2009a). Recycling batteries having basic electrolytes. US Patent 8,497,030.

Sloop, S.E. (2009b). Reintroduction of lithium into recycled battery materials. US Patent 8,846,225 B2

Sloop, S.E. (2016a). Reintroduction of lithium into recycled battery materials. US Patent 9,287,552 B2.

Sloop, S.E. (2016b). Reintroduction of lithium into recycled materials for battery. EU Patent 2,248,220 B1.

Sloop, S.E. (2017a). Relithiation in oxidizing conditions. US Patent 10,333,183.

Sloop, S.E. (2017b). Recycling positive-electrode material of a lithium-ion battery. US Patent 9,825,341 B2.

Sloop, S. E. (2017c). Recycling positive-electrode material of a lithium-ion battery. EU Patent 3,178,127 B1.

Sloop, S.E. (2018a). Relithiation in oxidizing conditions. Korean Patent No. 10-2018-7022741.

Sloop, S.E. (2018b). Relithiation in oxidizing conditions. Chinese Patent CN108886181A.

Sloop, S.E. (2019a). Relithiating in oxidizing conditions. US 2019-0273290 A1.

Sloop, S.E. (2019b). Method of recycling positive electrode material of a lithium-ion battery. Chinese Patent 106688135.

Sloop, S.E. (2019c). Recycling positive-electrode material of a lithium-ion battery. European Patent EP3178127B1.

Sloop, S.E. (2019d). Recycling positive-electrode material of a lithium-ion battery. Chinese Divisional Patent CN110676533A.

Sloop, S.E. and Allen, M. (2014). Recycling and reconditioning of battery electrode materials. US Patent 9,484,606 B1.

Sloop, S.E.; Crandon L.E. (11 November 2020) Battery Deactivation. US Patent 12,021,202.

Sloop, S.E., Kerr, J.B., and Kinoshita, K. (2003). The role of Li-ion battery electrolyte reactivity in performance decline and self-discharge. *Journal of Power Sources* 119–121.

Sloop, S., Crandon, L., Allen, M. et al. (2020). A direct recycling case study from a lithium-ion battery recall. *Sustainable Materials and Technologies* 25: e00152.

Swain, B. (2018). Cost effective recovery of lithium from lithium ion battery by reverse osmosis and precipitation: a perspective. *J. Chem. Technol. Biotechnol*, 93: 311–319. **https://doi.org/10.1002/jctb.5332**

US Department of Defense (5 April 2022). **https://www.defense.gov/News/Releases/Release/Article/2989973/defense-production-act-title-iii-presidential-determination-for-critical-materi/**

Vanderbruggen, A., Hayagan, N., Bachmann, K. et al. (2022). Lithium-ion battery recycling—influence of recycling processes on component liberation and flotation separation efficiency. *ACS ES&T Engineering 2* (11): 2130–2141.

Verzoni, A. (1 November 2024). **https://www.nfpa.org/news-blogs-and-articles/blogs/2024/11/01/missouri-battery-plant-fire**

Woyke, E. (2016). 24M: the startup's cheaper way to make lithium-ion batteries could make it cost-effective to store energy from renewable sources. *MIT Technology Review 119* (4): 76–80.

CHAPTER 6

Electric Vehicle Batteries: Repurposing for Second Life

Apoorva Roy
Department of Mechanical Engineering, University of Michigan,
Ann Arbor, MI, United States

Hamidreza Movahedi
Department of Mechanical Engineering, University of Michigan,
Ann Arbor, MI, United States

Anna Stefanopoulou
Department of Mechanical Engineering, University of Michigan,
Ann Arbor, MI, United States

Introduction

A key aspect of sustainable transportation energy is finding new purposes for electric vehicle (EV) batteries after they are no longer suitable for their original use in powering automobiles and before recycling. A battery's second life can take many forms, such as serving as an energy storage device or powering household appliances. Although this role is still emerging, we can learn a great deal about its direction by reviewing past and current applications.

What You Will Learn in This Chapter

In this chapter, the reader will learn about the:

- Battery life cycle; from in-use degradation leading to the end of its first life, followed by a second life wherein it can be reused, repurposed, refurbished, and/or recycled.

Electric Vehicle Batteries: From Sourcing to Second Life and Recycling, First Edition.
Edited by Bob Galyen and Frank Menchaca.
© 2025 John Wiley & Sons, Inc. Published 2025 by John Wiley & Sons, Inc.

- Repurposing and the value it adds to the most common EV battery chemistries.
- Different scenarios in which second-life batteries (SLBs) can be repurposed and the associated challenges.
- Safety concerns linked to SLBs and the liability conundrum that repurposing faces, along with industry standards that apply to battery repurposing facilities.
- Importance of diagnostic and prognostic techniques to enhance confidence in the utilization and performance of SLBs.
- Recommendations for policymakers, battery manufacturers, automotive original equipment manufacturers (OEMs), and SLB providers that can accelerate the proliferation of repurposed EV batteries.

Repurposing and Second Life

In batteries "second life" refers to the practice of continuing the usage of lithium-ion batteries (LIBs) at the end of an EV's lifespan before they are recycled. Figure 6.1 shows the battery life cycle, demonstrating that repurposing is one of the many ways to give batteries a second life. However, to fully appreciate the rationale behind repurposing, it is essential to understand the various components of a battery, their associated chemistries, and how they degrade over their lifetime. This section will cover all of these factors since they determine the viability of repurposing and help repurposers ascertain which application the SLB will be best suited for.

Second-Life Before Recycling

When an EV is deemed incapable of performing the drive cycles desired by its user due to battery deterioration (capacity fade and/or resistance growth), the battery pack is retired but can still be reused. This is because, for a good reliability reputation the automotive industry designs and builds EVs to last beyond the end of warranty (EoW) and drivers continue to use their vehicle or resell it even after the warranty period (Oak Ridge National Laboratory 2021). Warranty terms for EV battery packs in the United States suggest that an EV battery must retain more than 70% of its usable battery energy (UBE)[1] in eight years or 100,000 miles (whichever occurs first) for model years 2026–2030 and more than 75% retention in eight years or 100,000 miles (whichever occurs first) for model year 2031 onward (California Air Resources Board 2022a).

It is also projected that retired EV battery packs will be available in abundance. By 2030, around 1,000 GWh of retired batteries will be available worldwide, with China supplying 57% of the world's lithium battery waste and 75% of all EV batteries making their way into second-life applications (Colthorpe 2019).

---

[1] The definition and testing for UBE is discussed later in the section "Battery Degradation".

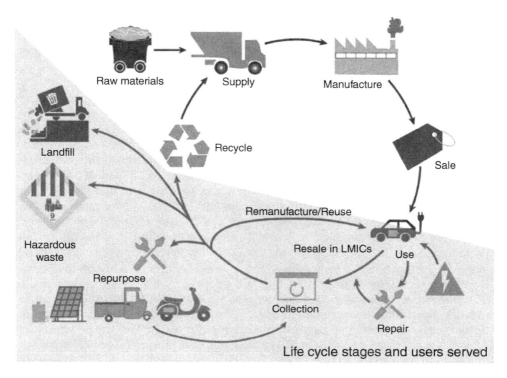

FIGURE 6.1 Life cycle of an EV battery and users served. The shaded region represents various second-life applications. *Source: Adapted from Kendall et al. (2024).*

The fate of retired batteries in their second life needs to be decided with caution in order to avoid wasting the tremendous number of resources that went into manufacturing them (Koroma et al. 2022). Recycling can help recover valuable metals such as nickel and cobalt, commonly found in EV batteries. Depending on the material recovery, recycling yesterday's battery could help build multiple future batteries since our manufacturing processes and pack assemblies will be leaner (reduced scrap, higher manufacturing efficiency, etc.). As technology matures to attain higher levels of standardization and material intensity improves, recycling will have to continue to be profitable. However, current recycling methods are complex, and the technology is still in its early stages, which can lead to high energy consumption and a net waste of energy if batteries are recycled immediately after being retired from EVs (Fan et al. 2023). Recycling also does not recover the embedded manufacturing value of used battery packs and most likely will face economic challenges as batteries start becoming increasingly reliant on more abundant and/or less expensive materials. Therefore, it makes sense to explore the economic and sustainability perspective of repurposing batteries in applications with less severe duty cycles – lower power and depth-of-discharge (DOD) requirements – compared to what they were subjected to in their first life. This is in line with the waste hierarchy that has been the guiding principle in the United States and the European Union (EU) to handle waste: *prevent, reduce, reuse, recycle, recover,* and *dispose*

(Environmental Protection Agency 2021). Repurposing also aligns with the U.S. Department of Energy's program for Catalyzing Innovative Research for Circular Use of Long-lived Advanced Rechargeables, or CIRCULAR (Advanced Research Projects Agency – Energy 2024). CIRCULAR envisions a circular domestic supply chain for EV batteries through regeneration, repair, reuse, and remanufacturing. It emphasizes the need for strategies that can prolong battery life and prevent premature recycling of functional materials.

Figure 6.2 depicts the various stages of EV use in the first life, by one or more users, followed by an eventual end of life, at which point a decision can be made on how to reuse the battery pack. Battery repurposing not only reduces the pressure to mine virgin materials but also promotes energy resilience (Zhu et al. 2024). Energy resilience means that communities and the grid can withstand and recover from power outages using battery energy storage systems (BESS) made of repurposed EV packs. Vehicle resale and eventual battery repurposing also offer an opportunity for EV buyers to recover some of their initial investment in the EV (since the battery pack forms the bulk of an EV's cost) via vehicle and pack resale (August 2023). In the long term, this can help advance EV adoption in the market. In addition to enhancing energy security and asset utilization, repurposing EV batteries also strengthens an economy's material security. Considering a mean EV lifetime of 15 years and various EV penetration scenarios proposed by the European Commission, European Network of Transmission System Operators for Electricity, and the International Energy Agency, Aguilar Lopez et al. (2024) stated that reusing even 40% of EV batteries can fully cover the EU's need for stationary storage by 2040 and make them less dependent on material imports.

Within an EV lithium-ion battery, the anode is typically made of pure graphite or graphite with small amounts of silicon, making the cathode the differentiating factor among various battery chemistries, particularly in terms of performance and cost. The cathode is made of lithium metal oxides containing precious metals that have limited reserves and are expensive to mine, leading to a growing interest in retrieving them through recycling methods (Baum et al. 2022; Whitlock 2024).

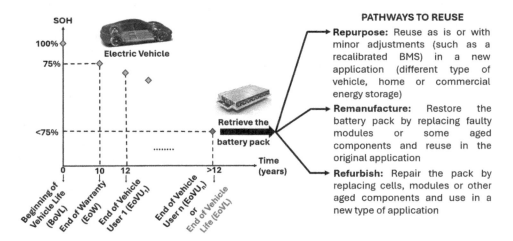

FIGURE 6.2 Simplified EV battery life cycle showing the various phases of EV resale (between users 1 and *n*) followed by the process of battery pack use in second life.

> **PRACTICAL INSIGHTS** | EV Battery Life Cycle and Pathways to Reuse
>
> Figure 6.2 shows the process of a vehicle with 100% state of health (SOH) at the beginning of life being used by one or more owners, even after it reaches the EoW period as defined by the OEM. Each time the vehicle's owner changes, the EV serves its user until the end of vehicle user n ($EoVU_n$) is reached, where n is an integer greater than or equal to one. If n = 1, it means that the first user has continued to use the EV, whereas n > 1 implies a resale has occurred. It is important to note that vehicle resale ensures a competitive total cost of ownership (TCO) for the previous user. When the EV reaches its ultimate End of Vehicle Life (EoVL), a repurposer retrieves the battery pack and decides how to utilize it by performing tests to evaluate its SOH and remaining useful life (RUL) post-first-life degradation. It is essential to do so in order to make a comparison among the potential second-life applications and choose the one that is the most economically viable, given the battery's capacity and impedance. For instance, if the pack has to be used in ESS, some functions can be more demanding (higher C-rate and DoD) than others. The various alternatives for second-life reuse include remanufacturing, repurposing, and refurbishing. Batteries with high RUL can directly be reused or remanufactured via minimal repairs. Battery packs typically consist of series-and-parallel-connected modules, with each module further consisting of series-and-parallel-connected cells. Packs have "weak" cells which can affect modules and, thus, remanufacturing might extend the majority of the pack's life. For those with low RUL, either the entire pack can be repurposed, or only the useful modules/cells can be salvaged and become parts for refurbished packs.
>
> While traditional repurposing methods involve SLB providers reusing entire packs for second-life applications, the prospect of disassembling them is valuable as it can increase the proportion of modules/cells in a retired pack that can be repurposed. Disassembly can also reduce the probability of a faulty or weak module/cell, making its way into a second-life use case it is not fit for. Repairing a retired EV pack can be challenging though, given its construction (the cells might be rigidly encased in addition to being welded together) and the hazard of fire or electric shock. To extend the lifespan of batteries, it is essential to design them in a way that ensures reparability while also creating an ecosystem of repair centers supported by a skilled workforce that is certified to work with high-voltage systems (Meegoda et al. 2024). Though the United States Environmental Protection Agency (US EPA) does encourage repair by exempting electronics repair, reuse, and repurposing-related activities from certain regulations, there is still tremendous scope for well-defined regulations specific to EV battery repair.

There is potential value in recycling Lithium–Nickel–Manganese–Cobalt (NMC) batteries as the constituent elements are expensive metals. The vast majority of recycling processes are expected to be profitable for cathodes containing high amounts of cobalt and nickel, which has led to the creation of many enterprises around NMC EV battery recycling (Chen et al. 2019). Globally, by 2050, 74% of retired NCX (cathode containing lithium nickel cobalt oxides, where "X" can be manganese or aluminum) packs will be repurposed, while the rest will be recycled (Xu et al. 2023).

Although NMC occupies more than half of the EV battery market share globally, Lithium–Iron–Phosphate (LFP) is gradually emerging as the chemistry of choice, accounting for 40% of the global market share in 2023 (International Energy

Agency 2024). While LFP batteries are currently concentrated in China and account for less than 10% of EV sales in the United States, they are projected to satisfy 20% of the demand in the United States by 2030 (Marjolin 2023). Their widespread availability and high-capacity retention (losing as little as 10% of their initial capacity in as many as 1000 cycles as per Severson et al. (2019)) make LFP batteries the perfect candidates for a second life, with repurposing being the best means to give them a second life and achieve a circular economy (Fallah and Fitzpatrick 2023). Moreover, the constituent metals in LFP cathodes, that is, lithium and iron, are of lower value which makes recycling neither necessary nor profitable. The low efficiency of lithium recovery via current state-of-the-art processes, such as hydrometallurgy, further makes LFP recycling unattractive (Wang et al. 2022). As shown in Figure 6.3, recycling LFP is currently not a profitable venture in most countries (Lander et al. 2021). However,

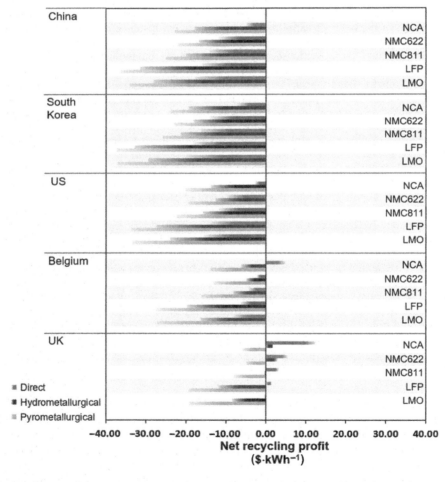

FIGURE 6.3 Net battery recycling profit ($ per kWh) for various recycling methods practiced currently and projected (such as directly recycling) and different battery chemistries in five countries (Lander et al. 2021), assuming transportation costs as given in EverBatt (Dai et al. 2019). EverBatt is an Excel-based battery recycling process and supply chain model developed by Argonne National Laboratory.

new technology by startups, such as ReElement Technologies, may recover lithium from LFP, making their recycling profitable (American Resources Corporation 2024). Meanwhile, most life cycle assessments (Slattery et al. 2024) conclude that repurposing should occur before recycling.

Battery repurposing is emerging as a preferred choice in other parts of the world as well, besides the United States and EU. In China, established battery recyclers such as GEM are repurposing retired packs into second-life applications like small power banks and utility vehicles (Janaky 2024). China's Ministry of Industry and Information Technology is also running a pilot program for recycling and second use of retired EV batteries in partnership with telecommunications giant China Tower, which will be collecting batteries from 16 major EV-makers across the country and using them to provide backup power to telecom towers (Jiao 2018). Innovative business models such as "Battery-as-a-Service (BaaS)" have also emerged, which involve collecting used batteries and utilizing their RUL before recycling by renting them to customers and providing after-sales services (AO 2021b).

Battery Degradation

Repurposers need to be aware of the limitations existing in battery packs retrieved from retired EVs due to in-use degradation during their first life. Degradation leads to a decrease in the stored energy and a drop in the power that can be drawn from the battery, ultimately leading to its inability to meet the performance requirement of the intended application "A." At that point, it is said to have reached end-of-life for application "A" (EoL_A) and can be repurposed into application "B" that would mark the beginning-of-life for application "B" (BoL_B). The most common mechanisms leading to capacity fade and resistance growth in batteries include lithium plating, particle fracture in electrodes, solid electrolyte interphase (SEI) layer formation and build-up, electrolyte depletion, and cathode transition metal dissolution (Wakihara and Yamamoto 2008). These mechanisms result in two degradation modes: loss of lithium inventory or LLI (lithium ions being consumed by undesirable side reactions) and the loss of active material at the negative and positive electrodes (LAM_{NE} and LAM_{PE}, respectively), which implies that electrode material is no longer available for lithium intercalation. Figure 6.4 shows the complex interplay between the causes of degradation, underlying mechanisms, degradation modes, and their ultimate effect on the battery.

All battery chemistries are prone to varying degrees of degradation or aging as a consequence of usage patterns, time spent in operation, and environmental conditions (especially during storage). Operating conditions lead to cycle aging, whereas time-dependent deterioration, which occurs even when the battery is not being used, leads to calendar aging. Cycle and calendar aging together, arising from a combination of physical and chemical phenomena, cause a reduction in the battery's stored capacity (represented by state of health – capacity or SOH-C) and a growth in the internal resistance of the battery (represented by state of health – resistance or SOH-R) (Plett 2015).

Although SOH is the most commonly used term to define battery health, its definition has not yet been standardized. For this reason, in the context of EVs, the United Nations Global Technical Regulation (UN GTR No. 22 2022) defines the state of certified energy (SOCE) and state of certified range (SOCR) as a percentage of the certified UBE or electric range remaining at a given point in time. UBE refers

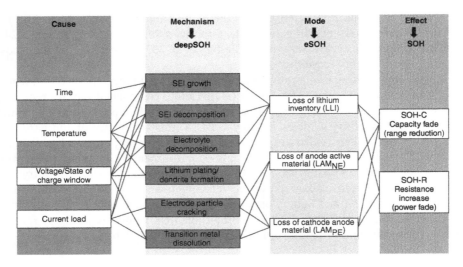

FIGURE 6.4 Causes of battery degradation and their corresponding mechanisms, modes, and effects in lithium-ion cells, observable in the form of deep-state-of-health (deepSOH), electrode SOH (eSOH), and SOH, respectively. *Source: Adapted from Birkl et al. (2017).*

to the total DC discharge energy measured in DC Wh using any of the standardized discharge test procedures developed by SAE (SAE J1634 2021). One of the test procedures is a full depletion test which measures UBE by driving an EV on a chassis dynamometer as per standard city and highway driving cycles until the battery is completely discharged. Another procedure involves recording the total energy discharged during a partial depletion test (dyno testing on city and highway drive cycles for a few hours) followed by discharging the remaining battery energy using a battery cycler (at a discharge rate in kW that is equivalent to driving at 65 mph). The latter procedure is gaining popularity as it reduces the amount of testing to be done on the chassis dynamometer.

While SOH-C and SOH-R are indicators of the battery's UBE and health as a whole, electrode-specific state of health or eSOH, which encompasses the electrodes' capacities and their utilization window, provides a deeper insight into battery health than LAM_{NE} and LAM_{PE} (Lee et al. 2020). This health metric, however, does not give us definitive information on the underlying dominant degradation mechanisms, which is why it is augmented with additional degradation states to define a term called deep-state-of-health or deepSOH (Movahedi et al. 2024). More details on eSOH and deepSOH will be provided in the section "Diagnostics and Prognostics to Promote SLBs."

Now let us assess if an EV battery pack at EoW would have sufficient range and energy to be resold or reused. Given that a new EV battery pack, on average, has a UBE of 72 kWh and a range of 380 km (Electric Vehicle Database 2024), at EoW (70% capacity retention) we can expect the aged pack to have a UBE of 50.4 kWh. We can conservatively consider that the battery's resistance will double, assuming linear correlation between capacity loss and resistance growth (Strange et al. 2021; Drallmeier et al. 2022). With an average energy consumption of 191 Wh/km, the used EV can provide a range of 264 km (164 miles). This range is sufficient to meet the demands of most drivers in the United States as 93% of vehicle trips are less than 30 miles (Federal Highway Administration 2022). In applications with low power requirements, such as residential power backup, the aged pack with 50.4 kWh can support the daily average energy demand of a household in the United States (U.S. Energy Information Administration 2023).

Pathways to Second Life

It is essential to understand the overlap and differentiation among the various terminologies used in the SLB industry, which are repurposing, refurbishing, and remanufacturing. Repurposing involves collecting old packs, selecting those with sufficient RUL for the new application, and combining them to form an energy storage system (ESS) that can be deployed in the second life. Refurbishing involves opening battery packs, replacing overly aged modules, and reassembling them to form a pack for second-life use. Refurbishing is more popular among SLB providers in Asia and Europe (Morris 2023). "Repurposing" and "refurbishing" are often used interchangeably, but the difference between them is that in the former, packs are "ready to use" while the latter requires some degree of dismantling and replacement of modules or a match-and-pair process that requires a large inventory of retired packs (Salza et al. 2021). While repurposing might require less time and investment, guaranteeing performance can be challenging if some modules are decaying faster than others. On the other hand, refurbishing can be time and labor intensive, and warranties on performance can be guaranteed only if the BMS is suitably equipped to perform its role in a second life. This is because retired EV packs are often disassembled to (at least) the module level, subjected to health assessment, and then reconfigured into a second-life application. Control and balancing capabilities are essential in a repurposed pack's BMS to successfully integrate and optimally utilize inhomogeneous retired packs with varying capacities. Integrating a new BMS with retired batteries is not expensive if done by adjusting the limits on power denials. Therefore, BMS repurposing can be done by keeping the BMS connections untouched but changing its parameters to reflect the faded capacity and increase in resistance for it to be able to communicate with the overarching master BMS of the second-life system (Börner et al. 2022).

Some SLB providers, such as U.S.-based B2U Storage Solutions, use a patented method to repurpose used packs from any OEM. Their approach employs a combination of controllers that interface with the BMS of the individual packs and can enable connection to a larger grid-connected ESS (Hall and Stern 2022). Others, such as Smartville Inc., seem to be using a similar approach to refurbishing heterogeneous battery packs, using custom power converters to achieve SOH balance (Smartville 2024). Cui et al. (2022) also addressed the challenge of power conversion when aggregating heterogeneous SLB packs with varying capacities and power limits that keep evolving as SOH drops. They proposed a lite-sparse hierarchical partial power processing approach, which minimizes the number of converters required, processes less power to achieve high system efficiency, and lowers costs.

A third alternative in second life is remanufacturing, which includes restoring an EV pack and reusing it in a vehicle (Salza et al. 2021). However, this could require time-consuming and expensive pack disassembly and can only be done if the majority of the retired pack has experienced minimal degradation (Nichols 2023). This approach works if a fault in the cells or cause of degradation in the modules is easily identifiable. Remanufacturing is beyond the scope of this chapter.

Potential Uses of Second-Life Batteries

Based on the application in which they are deployed, SLBs can be categorized according to their energy levels, purpose, and degree of mobility. High-energy SLBs are used in commercial applications, while lower-energy ones are used in residential applications. Table 6.1 provides a summary of the various applications in which SLBs can be deployed, given their respective energy demand, spanning from ESS to EV fast charging and small mobility vehicles. Based on their mobility, the second-life usage scenario can either be stationary (such as when integrated with wind or solar generation), quasi-stationary (providing energy at a mining or construction site), or mobile (for example, powering a forklift) (Hua et al. 2021). It is crucial to note that for all the aforementioned use cases, the battery's chemistry and the amount and types of degradation in the first life will determine the appropriate use case and the suitable environment for the battery in the second life.

TABLE 6.1 Viability of Second-Life Batteries in Various Applications, Given the Energy Demand.

Behind-the-meter Application	Residential	Commercial	Industrial
Renewables Integration	0–150 kWh	0–500 kWh	0–10 MWh
Energy Arbitrage	0–60 kWh	0–500 kWh	N/A
Peak Load Shaving	0–60 kWh	500–4500 kWh	0–4 MWh
Back-up Power	0–40 kWh	0–700 kWh	0–4 MWh
Small Mobility Vehicles	0–15.3 kWh	0–8 kWh	11–25 kWh
EV Charging	0–20 kWh	0–1 MWh	0–5 MWh
Demand Response	N/A	0–2 MWh	0–4 MWh
Microgrid	0.02–2 MWh	2–6 MWh	6–20 MWh

Capacities corresponding to each behind-the-meter application are specified under residential (homes and apartments), commercial (buildings meant for business activities), and industrial (structures built for manufacturing of goods and services) scenarios. SLB viability is represented by the color of the blocks containing the capacities. White indicates suitability, light gray implies they are sometimes suitable, and dark gray means SLBs are not suitable (Patel et al. 2024). Unsuitability (dark gray blocks) can be a result of economic infeasibility or very high C-rates.
Source: Patel et al. (2024)/Frontiers Media S.A/CC BY 4.0.

SLBs can be used to store power when the grid's generation exceeds the demand by integrating them with renewable energy sources such as wind and solar. Supply shortages and demand–supply mismatches in new LIBs may make retired batteries a viable alternative for ESS in the future. Xu et al. (2023) presented a conservative estimate in 2022 that, given the forecasts for EV battery deployment, market participation, and battery degradation, 30% of the global grid-storage demand will be met by second-life EV batteries by 2030. These ESS contribute toward enhancing grid reliability via a host of services such as energy arbitrage, peak shaving, and capacity reserve. SLBs are currently used mostly for short-term energy storage, such as 4-hour

ESS, which constitutes the bulk of the required storage capacity (Xu et al. 2023; Denholm et al. 2023). Depending on the application, the discharge duration of the ESS can range from 0.01 hours (for electric service power quality applications) to 4.5 hours for short-term storage (Neubauer and Pesaran 2011). Existing SLB ESS generally have a capacity of a few hundred kWh and are predominantly deployed in the commercial and industrial (C&I) segment. Repurposers are focusing on optimizing SL BESS integration with solar generation and EV charging, in addition to behind-the-meter applications like peak shaving.

Automotive manufacturers have been partnering with SLB providers to run pilot and commercial projects that can serve as a proof-of-concept for repurposing retired EV batteries. Recent notable efforts in the United States include Nissan, Tesla, Ford, and Chevy supplying 1,300 packs repurposed into a 28 MWh ESS by B2U storage solutions to be leveraged for grid services, and UC San Diego utilizing BMW's retired Mini-E packs as 100 kW/60 kWh renewable energy storage unit (Patel et al. 2024; Reinhardt et al. 2019). In other parts of the world, Renault, along with local SLB providers in the United Kingdom and France, has supplied residential and business ESS of the order of 50 MWh, composed of a mixture of new and old Renault Kangoo cells (Patel et al. 2024).

In some parts of the world, such as Japan and France (August 2023), retired batteries are also found in end-user applications as a power source for solar street lights, data centers, communication base systems, EV chargers, portable energy storage, home energy storage, camping recharge systems, and backup power for train signals (Dong et al. 2023). While China is also a large market for second-life EV batteries, IDTechEx reported in 2023 that "large-scale" BESS deployments containing repurposed batteries are unlikely to happen over the next few years. This is due to the short ban imposed by China's National Energy Administration between 2021 and 2022 citing concerns about the lack of systematic safety and health management systems (Shen 2021; Madalin 2023) for repurposed BESS and the recent over-supply of new batteries in China.

PRACTICAL INSIGHTS | Evaluating SLB Viability in a Given Application

Prior to deploying a retired battery pack in the second life, it is important to evaluate whether the battery's remaining capacity will support the intended application. When purchasing an off-the-shelf BESS made of retired batteries, users must confirm that the battery will not degrade within its warranty period. For *example*, if a manufacturer claims that the BESS will last for 3,000 cycles or 10 years, but the user intends to deploy it in bulk grid services which require daily charge–discharge (depending on the ESS sizing, it implies one full equivalent cycle or FEC per day) then the 3,000-cycle mark will be attained in *approximately* 8 years (because 365 cycles a year * 8 years = 2,920 cycles). However, if the BESS is going to be used for backup power, then it might last for 10 years as it most likely will not need to be cycled daily. It is also essential to distinguish between C-rate and cycle life when evaluating the performance of a battery. An EV battery is typically designed for 1,000 FEC over its lifetime (with an average range of 380 km on a single charge that corresponds to 380,000 km or 236,000 miles) and can withstand high charge/discharge rates (in the events of fast charging at 1.5–6C or accelerated driving). However, for an ESS while the C-rate might be low (C/4 in case of short-term storage) they are expected to last for more than 2,500 FEC over their lifetime. Therefore, it is important to consider the cycle life of a used EV pack being repurposed for energy storage.

Challenges in Second-Life Implementation

So far, we have established that battery utilization is maximized by enabling second-life use before recycling, and listed the potential pathways to facilitate reuse. However, there are roadblocks in the process of repurposing retired EV battery packs, which will be elaborated in this section. Some of them are as follows:

- Uncertainty about the volume of SLBs available and the variability among them.
- Integrating and controlling retired packs with inhomogeneous states of health while managing all the electro-thermo-mechanical aspects of operation.
- Difficulties in accessing confidential user data and BMS details to evaluate SLB health.
- Competing alternatives to repurposing (such as new installations and recycling).
- Making second-life use profitable, given the high risk of fast degradation and safety hazards along with the liability burdens.

Module Selection and Adequacy

Electric vehicle battery packs are composed of a number of modules that may degrade differently during the first life and consequently have different capacities at the EoL. However, battery management systems typically report pack-level SOH but provide little information on the pack's inhomogeneity, durability, and future performance capability, that is, RUL for the new application. Moreover, the pack's SOH typically captures the SOH of the cells with low capacity or high internal resistance, also termed as the "weak" cells. As per Drallmeier et al. (2022), it is important to note that some weak cells may burden the discharge or charge utility of the entire pack, as the weak cells hit the minimum or maximum voltage limits, respectively. To protect against unsafe over-discharge or overcharge the first-life BMS stops the operation of the entire string of cells in series, therefore limiting the utilization of the "strong" cells. Tracking the SOH of the strong cells in a pack will help determine how many and which modules can be repurposed to be used as ESS (Lorscheid 2024). A pack consisting of a single weak module can be repaired and refurbished/repurposed if the weak module is replaced. Alternatively, a weak pack may contain some perfectly healthy modules that can be salvaged and refurbished. Having enough modules is also critical to ensure the smooth replacement of faulty/depleted modules and prevent downtime.

Extent of Reutilization

The two main models of battery pack reutilization are "teardown" and "whole-pack" (Morris 2023). The teardown model is cumbersome to implement as it involves disassembling the pack and introducing new parts/algorithms which is labor-intensive, expensive, and creates room for defects. Moreover, most modules within modern packs have large form factors (such as BYD's "blade battery") and are integrated into a cell-to-pack design (pioneered by CATL) which makes teardown even more difficult (AO 2022b). Therefore, repurposing the whole pack is gradually becoming the business paradigm. However, successful integration of a repurposed pack with the grid is subject to quality control, pack characterization, connecting the communication and thermal management cables, and developing a second-life BMS that controls power flow between various packs and also coordinates with a master ESS controller (B2U Energy 2024).

Aged packs also come with increased inhomogeneity between individual cells, parallel groups of cells, and among modules, as illustrated in panels A and B of Figure 6.5a. Variability in charge/discharge capacity increases with aging, as shown in Figure 6.5b, wherein individual cells are cycled within the same voltage range. When cells are connected in series small variability in early life diverges and grows even more than individually cycled cells (Reniers, Howey 2023). Parallel connections, however, tend to even-out cell-to-cell variability (Weng et al. 2024). If the imbalance between

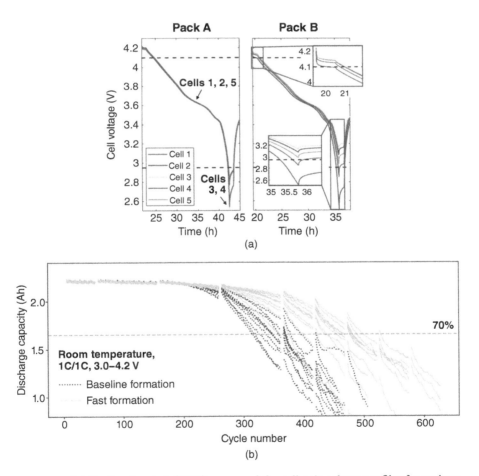

FIGURE 6.5 (a) Drallmeier et al. (2022) compared the cell pair voltage profiles for various packs (two of which are shown in this figure) retrieved from discarded DeWALT power tools and showed that even small imbalances in the cell pairs lead to changes in cell pair voltages at low SOC. This can result in a large difference in the capacity of cells when operating at the minimum terminal voltage. *Source: Drallmeier et al. (2022)/with permission of Elsevier.* (b) Variability in the discharge capacity grows as cells age during a 1C/1C cycling test at room temperature. *Source: Weng et al. (2021)/with permission of Elsevier.*

cells in a pack has not been addressed in the battery's first life, it definitely needs to be addressed in the second life. This is because one should not consider repurposing modules with high amounts of imbalance among cells in series as they can lead to faster capacity fade (ReJoule Energy 2021). While cell balancing strategies do offer a potential solution, their effectiveness depends on the usage pattern after repurposing.

State of the Market and Technology

From a techno-economic standpoint, developing business models that evaluate the economic viability of repurposing is a work in progress, since battery and energy prices are evolving rapidly, and repurposing is profitable only under certain market conditions. In a rapidly advancing battery technology landscape, reusing SLBs for ESS presents a major challenge. By the time these batteries reach the second-use stage, their technology is often outdated (at least a decade old). To avoid compatibility issues when integrating them with modern ESS, a master BMS is required to manage heterogeneous packs (Meegoda et al. 2024).

SOH and RUL Estimation

During first life, the BMS is required to accurately estimate batteries' SOH in order to optimize their utilization and ensure safety. While a plethora of literature exists on online estimation of SOH (i.e. during operation), these algorithms are designed using deep depth of discharge and calibrated to work under a specific set of operating conditions. Moreover, the BMS is also expected to detect and predict rapid capacity fade (also known as "knees," more details in the section "Diagnostics and Prognostics to Promote SLBs"), which is a challenging task. A new second-life BMS (SL BMS) should be designed based on the use case expected in second life. A reliable SL BMS of this sort may help SLB batteries to last for a longer amount of time in a second-life application, although the total economic benefit from derating for longer life needs to take into account the specific SLB use and cost structure (Kumtepeli et al. 2024). If equipped with predictive capabilities, SL BMS can also optimize battery dispatch such that the knee point is delayed or avoided.

The time, distance traveled, or energy throughput until a battery's SOH falls below a minimum threshold in its first life is called remaining useful life or RUL (Weng et al. 2023), and it is a key metric determining EV and ESS warranty terms. Estimating how much of the battery's initial capacity is still available for use in second life is a difficult and time-consuming process but extremely valuable in helping SLB providers make an informed decision on how to handle retired batteries. Panels (a), (b), and (c) in Figure 6.6 demonstrate linear, self-limiting, and accelerating aging trajectories, respectively, while panels (d), (e), and (f) illustrate how the knowledge of RUL can determine the choice between recycling and reselling an EV battery. Panels (d) and (e) are where the EV has the potential to be resold and recycled since the battery has sufficient RUL post the 70% SOH milestone, whereas in panel (f), recycling is the only option since the battery's RUL is very short. Due to the importance of knowledge of RUL for SLBs being deployed in BESS, the U.S. Department of Energy mandates BESS to have a service life of at least 10 years since commissioning (U.S. Department of Energy 2024). Several companies, national labs, and academics have devised techniques to rapidly determine the SOH-C/R of retired batteries after unknown usage

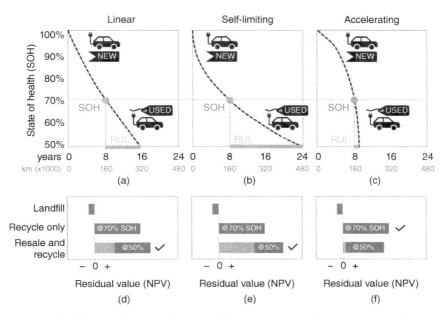

FIGURE 6.6 (a), (b), and (c) denote linear, self-limiting, and accelerating aging trajectories, respectively. In all the cases, the SOH markers denote the EV battery meeting its warranty terms (8 years, which more-or-less corresponds to 70% battery SOH or 30% degradation). However, due to different aging trajectories in second-life they have a wide variation in their residual values as shown in (d), (e), and (f). *Source: Weng et al. (2023)/with permission of Elsevier.*

scenarios. One of the first diagnostic tests developed to aid this process was the hybrid pulse power characterization (HPPC) protocol, first developed by the United States Advanced Battery Consortium LLC in 1996 (Idaho National Laboratory 1996) and most recently revised in 2015 (Christopherson 2015). As shown in Figure 6.7, HPPC is a reference performance test (RPT) procedure that is used to determine the pulse power and energy capability under no-load conditions as a function of aging.

Others such as Zhang et al. (2020) and Gasper et al. (2022) have used a combination of electrochemical impedance spectroscopy (EIS) measurements taken at different states of charge and temperatures along with machine-learning models like Gaussian Process Regression to forecast SOH and RUL. Their algorithms are able to predict RUL within reasonable error bounds in unforeseen second-life scenarios while using limited historical data from first life.

Accurate SOH and RUL estimation have far-reaching commercial benefits for SLB providers. SLBs mostly make their way into ESS which are not an asset class that insurance companies are well-versed in. In such scenarios, SLB providers and BESS operators can use predictive diagnostics to forecast SOH and RUL, flag hazards, and enhance battery safety. This, in turn, can help secure more favorable insurance terms (Energy Storage News 2024; ACCURE Battery Intelligence 2024).

Economic Viability and Price Limitations

The economic advantages of using batteries in second life rely heavily on the cost of the battery itself, its residual performance, overheads associated with shipping

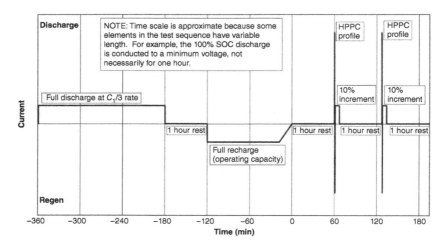

FIGURE 6.7 Hybrid pulse power characterization test (start of test sequence). The 10% increments continue until the battery is discharged.

and handling, any intervention for re-manufacturing, integration with other heterogeneous assets, and how much one can save on electricity bills by using them for energy storage. To justify second-life use, repurposing should be more profitable than recycling, the cost of the SLB should be proportional to its RUL, and the total cost of procuring and refurbishing the retired battery must be much lower than that of purchasing a new one. Open-source tools such as the National Renewable Energy Laboratory's Battery Second-Use Repurposing Cost Calculator can be used to understand when and in what condition used EV batteries will become available, how much it will cost to repurpose them, taking into account the auxiliaries, and how long the repurposed batteries will last in their second life (National Renewable Energy Laboratory 2020).

A concerning projection in battery prices is a plateau in the price of SLBs and a decline in their price advantage with respect to new batteries, owing to the rapid decrease in prices of new batteries (Kamath et al. 2020; Sun et al. 2018). If an SLB provider decides to use the batteries for energy storage, they need to ensure that the battery will have sufficient capacity and cycle life to last through the payback period, that is, the time it will take to recover the initial investment on the project. The reliability and accuracy with which they can estimate SOH and RUL will play a key role in this aspect. More details on SOH and RUL estimation will be provided in the section "Diagnostics and Prognostics to Promote SLBs".

There are two price limits that determine the economic viability of SLBs. The lower limit is the "willing to sell" price, which is the price at which the owner of a used battery is willing to sell it. The upper limit is the "market evaluation" price, which is the highest price at which the market will buy an SLB, depending on its SOH, RUL, repurposing costs, battery chemistry, and the price of a new battery. The market price for SLBs will be competitive only when the "willing to sell" price is lower than the "market evaluation" price (Wu et al. 2020). Bach et al. (2024) projected that the fair market value of retired EV LFP batteries is 40% higher than the market value of a new battery and presented a strong case for repurposing LFP in applications with daily cycles. Repurposing LFP is a better option in the United States than it is in China because new batteries are traded at higher market prices in the United States. On the contrary, there was only a marginal economic case for repurposing NCX batteries in the United States (Bach et al. 2024).

In addition to the above, the levelized cost of energy (LCOE), which is the minimum cost at which an ESS operator must sell electricity to break even over the course of operating a BESS, must be competitive enough to justify using a repurposed or refurbished battery pack for energy storage. It is calculated by dividing the average total cost to build and operate an asset by the total energy output of the asset over its lifetime. Finally, a positive net present value (NPV) is also a prerequisite to warrant an investment in SLBs (Dong et al. 2023). The present value of a future series of payments and revenues that will occur over the entire life of an investment is known as NPV. A positive NPV indicates that the investment will be profitable. Depending on the load profile and the segment in which it is being used, the NPV of a BESS made of SLBs can vary significantly. For example, as per an analysis by Bai et al. (2019), for most regions in China, rooftop solar paired with retired EV batteries yielded negative NPV in residential applications but positive NPV in the commercial/industrial sector, owing to the variation in their energy demand.

Safety

Common battery safety issues – which can get exacerbated with age – include thermal runaway due to uncontrolled heat generation, over-charge/discharge, short circuit due to electric abuse, mechanical damage, and thermal shocks (Chen et al. 2021). Safety hazards, though rare in first life (approximately 1 per 10,000 vehicles as per Huang et al. (2021)), have raised concerns over battery deployment in both first and second lives. There are currently no known cases of fires occurring in BESS-containing SLBs. However, there is a concern, given that fires occurred in the past. According to the BESS Failure Incident Database, the overall rate of fires in first-life BESS has sharply declined by 97%, with the failure rate dropping from 9.2 to 0.2 failures per deployed GW (#/GW) between 2018 and 2023.

Fault diagnosis can play a major role in ensuring the safe operation of batteries, wherein the BMS monitors cell groups and modules to identify parameters like overcurrent and overvoltage. The BMS can report state-of-safety (SOS), which in the research community has been defined as the inverse of "abuse" by Jossen et al. (Cabrera-Castillo et al. 2016). It is a binary metric indicating whether a battery is safe to use. For an EV battery and its BMS that has been certified by an OEM, SOS = 1 which means that the battery can be used until EOL without causing any damage to life or property when operated as per manufacturer recommendations within the warranty period in the application that it is intended for. However, when a battery is being repurposed for second-life use in an unforeseen scenario, the SOS drops to 0. It should be the repurposer's responsibility to test the pack as per UL 1974 and recalibrate the BMS such that SOS gets reset to 1. Resetting SOS to 1 should be associated with the transfer of liability.

In this section, we will summarize the industrial standards that repurposed batteries must comply with and highlight the complexities and gaps in state-of-the-art methods for assessing and certifying battery safety. We will also address the misconceptions prevailing around the safety of repurposed BESS units.

Industry Standards

U.S.-based auditing leader Underwriters Laboratory (UL) in 2018 published the world's first certification for retired EV batteries, namely UL 1974: the Standard for EV Battery Repurposing Facilities (Underwriters Laboratory 2019). It is the U.S. and Canada's binational standard for sorting/grading packs, modules, and cells, determining SOH, and evaluating the potential for being used as energy storage in second life. Having a UL 1974 certification, although not a mandatory requirement in North America, certainly eases the process of getting ESS deployment permits from local authorities for SLB providers. It encompasses a broad range of activities from visual inspection of SLB samples, analysis of BMS data, disassembly of packs, grading batteries for repurposing, quality control, tracking the storage condition, and disposal of rejected parts. One of the pioneers in developing a standardized process to grade EV battery packs for second-life use is 4R Energy Corp., which is a joint venture between Nissan and Sumitomo Corporation (Patel et al. 2024). Their methodology is as follows:

- **Grade A**: Minimal degradation; battery can be reused in a new EV.
- **Grade B**: Moderate degradation; battery can be repurposed into an application such as stationary energy storage.
- **Grade C**: Substantial amount of degradation; battery can be used for backup power.
- **Grade D**: Battery cannot be repurposed and should be recycled.

Energy storage units composed of repurposed batteries must also comply with the standards pertaining to commercial BESS. One of them is UL 9540 (Bashevkin 2022), which is the Standard for the safety of energy storage systems and equipment and is used to certify the safety of ESS at the system level. It must be noted that obtaining UL 9540 certification is not possible without destructive testing of representative samples and factory product certification (no field testing). To facilitate that, UL 9540A specifies the test procedure for assessing thermal runaway and fire propagation in ESS at the module and unit level. Thermal runaway is defined in NFPA 855 (the Standard for the Installation of Stationary Energy Storage Systems) as the uncontrollable rise in a battery's temperature when it is unable to dissipate the heat being generated, but which does not necessarily lead to venting, fire, and/or explosion. Performing the UL 9540A procedure produces a test report that a Registered Design Professional and Authority Having Jurisdiction will use to determine acceptability. UL 9540A is, therefore, a test method that helps manufacturers understand what happens when an ESS undergoes propagating thermal runaway. Stakeholders analyze the test results and decide if the system performed as expected. Authorities Having Jurisdiction interpret the test results to determine compliance with codes and standards and approve installations.* Finally, the batteries must also fulfill the requirements of UL 1973 (referenced within UL 9540) to ensure the safety of personnel operating the ESS in a real-world scenario. In conclusion, UL9540 and the standards referenced therein (UL9540A and UL1973) do not directly apply to SLBs but do come into effect when retired EV batteries go into their most common second-life use, that is, ESS.

While UL and the EU Battery Regulation do provide some safety measures for retired EV batteries, regulation in this area is still a work in progress. In light of this situation, some BESS developers such as Ecostor and testing services providers such

---

*Source: Communication with Alex Schraiber, UL Solutions.

as National Instruments have developed their own rigorous safety standards for SLBs (ECO STOR 2023; National Instruments 2024). Automobile manufacturers can also help streamline the SLB safety certification process by complying with UL 2580 (Testing Standard for Batteries for Use in Electric Vehicles, which also contains fire propagation containment requirements) and sharing critical BMS and testing data with repurposers.

Safety of Aged Cells

Battery safety assessment is usually done with "unused" or "uncycled" cells, but cells are unused for only a small fraction of their lifetime. Safety test reports issued as per UL standards only account for tests done on uncycled cells and failure mitigation mechanisms are based on the results of these tests (Preger 2024). This implies that there is a lack of testing for aged cells as per existing standards. It is essential to understand how the thermal, mechanical, and electrical abuse of aged cells varies from that of uncycled cells so that effective failure mitigation mechanisms can be devised for second-life ESS systems. Experimental investigation into the SOS of aged cells is necessary as they exhibit unique responses based on their chemistry, aging level, size, form factor, and abuse protocol. Whether or not an aged cell is safer than an uncycled cell depends on the cell chemistry, evaluation metric, and the aging protocol.

Preger (2024) studied the thermal, electrical, and mechanical abuse response of lithium-ion cells that were aged by different protocols and identified the critical parameters that can be used to evaluate the safety of aged cells relative to fresh ones. When subjected to thermal abuse, an aged cell is considered safer than its fresh counterpart if it has a higher thermal runaway onset temperature or attains a lower peak temperature. From a degradation point of view, aging cells under conditions that cause lithium plating (such as charging in cold climates) renders them less safe due to a reduction in self-heating and thermal runaway onset temperatures by over 50 °C, along with an increase in heat released from the reaction of plated lithium. Electrical abuse tests involve over-charging, over-discharging, and external shorting of cells to assess thermal runaway behavior (incidents of venting, fire, and/or smoke) and the ability of the battery's internal safety devices (such as the current interrupt device or CID) to prevent it, in addition to temperature response. SLB management systems must include tolerance for chemistry-specific over-charge and over-discharge scenarios. Mechanical abuse of aged cells involves tests such as nail penetration, crush, indentation, pinch, three-point bend, and lateral compression. Kovachev et al. (2020) showed that aged cells required greater mechanical intrusion to induce an internal short and are less violent than fresh cells during thermal runaway because of lesser stored capacity.

While the studies mentioned above offer useful insights into the safety of aged cells, literature on this subject is still limited in terms of the number of studies and diversity of cell chemistries, age, and sizes considered. Thermal abuse protocols should also account for the cell's thermal environment and exposure to in-field shock and vibrations, as BESS composed of aged cells can be deployed in diverse locations and climates. As for electrical and mechanical abuse tests, the results are not statistically significant enough to draw firm conclusions. Due to inherent variability in SOH across a battery pack, it is essential to go beyond cells and characterize safety at the module and pack levels (Preger et al. 2022), especially because whole-pack repurposing is the preferred choice among SLB providers. There also needs to be a greater focus on studying LFP cathodes and early indicators of impending fire hazards such as off-gassing events, that is, when the cell starts venting prior to thermal runaway (Preger 2024).

> **PRACTICAL INSIGHTS | Certifying SLB Safety**
>
> Figure 6.8 encapsulates the host of regulations that repurposed BESS units need to comply with. National Fire Protection Association (NFPA) 855 and International Fire Code (IFC), which outline the requirements for fire protection and safe installation of modern BESS units, call for BESS to be certified to UL 9540. UL 9540 is the currently most evolved standard for the safety of BESS and in turn requires batteries to be certified to UL 1973, which applies to batteries for use in stationary, vehicle auxiliary power, and light electric rail (LER) applications. Ultimately, all BESS composed of retired batteries must be tested and validated by a repurposing facility that has received UL 1974 certification, which is the standard to be upheld by EV battery repurposing facilities.
>
>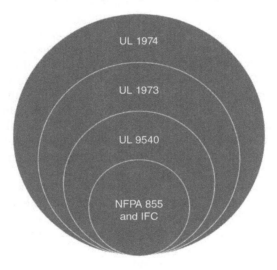
>
> **FIGURE 6.8** Regulatory framework (in hierarchical order) for second-life batteries. NFPA 855: Standard for the Installation of Stationary Energy Storage Systems, IFC: International Fire Code, UL 9540: Standard for Safety – Energy Storage Systems and Equipment, UL 1973: Standard for Safety – Standard for Batteries for Use in Stationary, Vehicle Auxiliary Power and Light Electric Rail (LER) Applications, UL 1974: Standard for Safety – Evaluation for Repurposing or Remanufacturing Batteries

Safety Hazards: Concerns and Misconceptions

Irrespective of the domain in which retired batteries will be utilized, the liability of these batteries remains a lingering issue since the second-life market is largely unregulated (Börner et al. 2022). In the event of an accident, it is hard to trace the battery back to its original manufacturer (Meegoda et al. 2024). Fire hazards and the risk of additional liability are major reasons behind OEMs being hesitant to build an SLB ecosystem (Zhu et al. 2021). Safety hazards associated with batteries also make their transportation for sorting, storage, and match-for-remanufacturing a challenging, and thus, expensive task. LIB shipments are categorized as Class 9 ("Miscellaneous") hazardous materials by the U.S. Department of Transportation (DOT) and must be packaged in a way that precludes the possibility of accidental fires, short circuits, and/or mechanical damage.

Due to a lack of knowledge about the battery's SOS at the end of its first life, repurposers often adopt a conservative approach and underutilize the battery in its second life in an attempt to reduce the risk of safety violations (Niese et al. 2020). However, there is not enough data to correlate the failure of BESS systems with age. Insights from EPRI's BESS Failure Incident Database (Srinivasan et al. 2024) show that only a small fraction of failure incidents occurred in systems older than three to four years (Figure 6.9a). Moreover, they are overall system failures and not reflecting battery components alone. Srinivasan et al. (2024) identified the root cause of BESS failures for 32% of the total number of failures and showed that most of them occurred during the integration, assembly, and construction phase. An important conclusion, shown in Figure 6.9b (Srinivasan et al. 2024), was that 43% of the failed elements were in the balance of system (BOS) and another 46% were in controls. BOS refers to components except the cells, modules, and controls, typically consisting of but not limited to busbars, cabling, power conversion systems (PCSs), transformers, and liquid cooling systems. Control systems coordinate the operation of the ESS, including the BMS, energy management system (EMS), plant controllers, and any subsystems. Notably, most of the failures attributed to controls were based on investigations that found accidents happening when cells were in high SOC range and attributed the failure to aggressive operational limits set by the control unit.

As rapid growth in the deployment of ESS is expected, experts at EPRI anticipate a "bathtub" curve for a component failure rate over life. This means that the rate would be high at beginning-of-life, followed by a decline around middle-of-life, and potentially increase toward end-of-life. In a sense, Figure 6.9a represents only the first half of this "bathtub" trend.

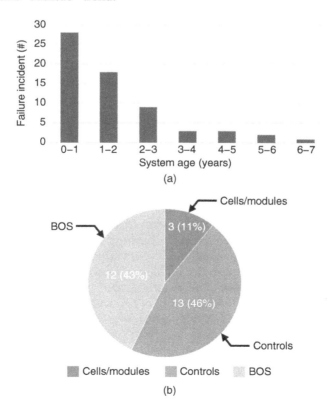

FIGURE 6.9 (a) BESS age at failure, where known (b) breakdown of BESS failures by failed element. *Source: Srinivasan et al. (2024)/U.S. Department of Energy/Public domain.*

Diagnostics and Prognostics to Promote SLBs

SLB providers may not necessarily be aware of how the battery was used in its first life, even if they can test the retired pack for SOH-C/R. Depending on how they were used in their first life, retired EV battery packs can have a wide range of remaining useful capacities in the second life. Accordingly, different applications with varying intensities of use and environmental conditions would be suitable for batteries with different levels of degradation. Using batteries that may be unfit for the intended application poses serious financial risks, which is a major concern for OEMs, SLB providers, and BESS operators. Specifically, for second-life applications to be economically viable, the batteries need to last for a certain amount of time so that repurposers can enjoy returns on the capital invested toward repurposing them. In addition, it is also typical for batteries to experience rapid capacity fade, also known as "rollover" or "knee" point (Attia et al. 2022), toward or after the end of their first life at 70–80% SOH. Since batteries are typically repurposed after hitting the 70–80% SOH mark, it is more likely for them to experience knees.

The key to resolving all of the aforementioned issues lies in a thorough assessment of the battery's health, knee-point prediction, and an accurate prognosis of its RUL at the end of its first life. Such robust diagnostics and prognostics serve as both a safety measure and a tool to reduce the cost of future maintenance and downtime. In conclusion, the benefits of accurate SOH quantification, SOS assessment, and RUL estimation are twofold: (1) assessing the economic viability and SOS of the battery packs before the second life; (2) online estimation of SOH during the second life.

Estimation and Prognostics for Assessing the Pack for SL

Estimating the battery's health and predicting its RUL accurately is a critical and challenging task in both the first and the second life. While the United Nations Global Technical Regulation (GTR) and the California Air Resources Board (CARB) define an eight-year warranty for EV batteries, they do not specify how RUL should be quantified in the second life. It has also been mandated by CARB that the SOH be reported with 5% accuracy in the state of California by 2026 (California Air Resources Board 2022b). However, Weng et al. (2023) showed that 5% measurement uncertainty in SOH leads to a significant decline in the confidence level of RUL predictions. Moreover, a single SOH measurement taken at the end of the battery's first life cannot be used to predict RUL because packs having the same SOH can have different underlying aging trajectories (owing to their chemistry or duty cycle in the first life), and consequently last for different durations in the second life. This phenomenon, also known as "path dependency of degradation" (Dubarry et al. 2012), is illustrated by examples shown in panels (a)–(c) in Figure 6.6, which have linear, self-limiting, and accelerating trajectories, respectively. The trajectories correspond to very different second lives even though all of them had the same SOH-C (70%) at the end of their first life (8 years). In the context of this figure, RUL refers to the years between 70% and 50% SOH, and it can be observed that the self-limiting scenario yields the maximum RUL and would be the best choice for second-life use.

The issue of lack of historical data on EV battery SOH is addressed in CARB's proposed regulation "Order for Zero Emission Vehicles" (ZEV) for 2026 and subsequent model year passenger cars and light-duty trucks (California Air Resources Board 2022b). As part of ZEV in-use verification reporting, manufacturers will be required to report SOH at two distinct points during the EV's useful life – first, when the vehicle has been in service for more than three years (driven more than 36,000 miles) and second when the vehicle has been in service for more than six years (driven more than 60,000 but less than 150,000 miles).

Conventional SOH testing methods for EV batteries, such as slow charge–discharge tests, are laborious and expensive and hence not scalable for the large volume of batteries that need to be assessed before being deployed in the second life. Therefore, SLB providers use various techniques to speed up and streamline the testing process. Similar to the approach adopted by Drallmeier et al. (2022), ReJoule Energy uses AC impedance data gathered from EIS across a range of frequencies to estimate battery SOH (ReJoule Energy 2024c). They build and validate models for SOH estimation such that the SOH estimated using impedance data is benchmarked against the SOH measured via a cycle test (ReJoule Energy 2024a). These models can later be used to predict SOH without subjecting batteries to prolonged cycling. As a diagnostic method, EIS is beneficial because it is noninvasive, reliable, reduces testing time, and can be done on a high-voltage EV pack directly through the charging port (ReJoule Energy 2024b). However, as highlighted by Drallmeier et al. (2022) who tested discarded DeWALT lithium-ion power tool battery packs at random, models that estimate SOH based on the correlation between capacity and resistance are robust only when built after extensive testing. Moreover, such models vary with battery chemistry, form factors, and vendors as shown in Figure 6.10b, which makes it obligatory for those assessing SOH to create databases for different types of batteries originating from different sources and production batches.

Some SLB providers, such as RePurpose Energy, Inc., are leveraging data-driven methods like machine learning to test, sort, and decide whether to recycle or repurpose EV batteries and find suitable second-life applications (Advanced Materials Manufacturing Technologies Office 2024). It is fair to assume that data-driven methods rely on coordination between the repurposer and the provider of retired batteries, as it is essential for the repurposer to have access to BMS data to train algorithms that can estimate SOH and predict RUL under second-life usage scenarios.

PRACTICAL INSIGHTS | Resistance or Fast EIS as a Surrogate for SOH

Drallmeier et al. (2022) developed a method to perform rapid initial screening of SLBs by computing ohmic resistance via charge and discharge interrupt cycles (by dividing instantaneous change in voltage by change in current) and showing that the charge-interrupt internal resistance was correlated with internal capacity. Their test protocol is shown in Figure 6.10a, where it can be seen that the three-hour charge/discharge interrupt tests are much shorter than the standard 30-hour HPPC tests. Figure 6.10b depicts how the relationship between capacity and $R_{s,CI}$ (internal ohmic resistance calculated

(continued)

> **PRACTICAL INSIGHTS** | Resistance or Fast EIS as a Surrogate for SOH (*continued*)

via a charge interrupt) can be approximated by a linear fit. Among a group of discarded packs collected from a DeWALT power tool labeled from A to M (Drallmeier et al. 2022), pack H has the highest resistance, which translates to the lowest measured capacity. Packs E, L, and N have some of the lowest resistances and highest measured capacities (more than that of the fresh pack A). Pack M, however, has the second lowest capacity despite its resistance being almost equal to that of the fresh pack A. Packs M and D limit the linear relationship between capacity and resistance because tests on individual cells clarified that there was high variability in resistance and capacity among them. As a corollary, one can hypothesize that inhomogeneity in the aging of cells connected in series and parallel defies simple EIS vs. SOH-C correlations. The linear relationship between $R_{s,CI}$ and capacity can be an indication of LLI caused by the consumption of lithium ions by the SEI layer when the cell is cycled (Prasad and Rahn (2013)). As the thickness of the SEI layer increases, the resistance of the cell grows (Plett 2015; Pannala et al. 2024). An added advantage of using $R_{s,CI}$ as a capacity metric, as per Drallmeier et al. (2022), is that it remains constant for mid- to high-range terminal voltage values, eliminating the need to know the exact SOC to estimate SOH.

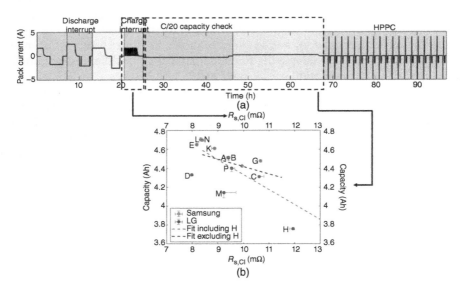

FIGURE 6.10 (a) Pack current profiles used to characterize retired battery packs (Drallmeier et al. 2022), consisting of a discharge interrupt (C/3 charge and C/2 discharge rate), charge interrupt (C/3 charge with interrupt), and HPPC (hybrid pulse power characterization which involves a series of 10-minute C/3.33 constant-current discharge pulses, followed by a 60-minute rest period). (b) Linear correlation between capacity and corresponding ohmic resistance calculated via a charge interrupt ($R_{s,CI}$) for a pair of cells in 13 packs (A–M). The bars associated with each point denote the range of resistance and capacity associated with a total of 13 packs with 6 from one vendor and the rest from another. Pack H has a large cell capacity imbalance that clearly created an outlier from the general linear relationship. *Source: Drallmeier et al. (2022)/with permission of Elsevier.*

Path Dependency of Degradation

Degradation in LIBs is usually quantified via aging models that are parameterized using experimental data collected by cycling batteries under various stress factors such as temperature, charge/discharge rates, and depth of discharge. Generally, experiments are performed by keeping the stress factors fixed throughout the charge–discharge cycles. However, when transitioning from the first to the second life, these operating conditions might not remain constant. For instance, in the United States, retired EV batteries from a vehicle being driven in Arizona might make their way into an ESS being deployed in Michigan, implying that the battery will experience a significantly colder climate in its the second life. To be precise, in Arizona (hot weather), the dominant degradation mechanism would be SEI layer growth, whereas in Michigan (cold weather), it could be lithium plating. Therefore, aging models must take such path-dependent degradation into account since noncommutative capacity fade is observed when the order in which the stress factors are applied is changed (Karger et al. 2022). Roy et al. (2023) developed empirical models for cycle and calendar aging in both hot and cold conditions. They then developed switching methodologies to predict degradation for switching from hot-to-cold cycling and vice versa within a 5% root mean square error. Such data-driven and empirical models can achieve limited success in predicting degradation under changing operating conditions, as they might not be able to extrapolate under conditions which they have not been trained under.

DeepSOH to the Rescue

Figure 6.7 illustrates an example of how path dependency affects the prediction of battery degradation. The RUL of batteries undergoing nonlinear (self-limiting and accelerating) degradation is substantially different from the one degrading linearly. RUL is also dependent on the SOH at the end of first life. Figure 6.11 addresses the case when the duty cycle changes in the second life, RUL can vary substantially depending on the severity of use, and predicting an accurate RUL becomes even more challenging. In such scenarios, higher-level SOH metrics such as capacity, resistance, and electrode-level SOH, though necessary, are not sufficient to predict RUL. A digital twin of battery degradation was used to demonstrate that cells with identical SOH-C and SOH-R may degrade at different rates during the second life when subjected to the same duty cycle. This is due to various internal states associated with mechanical degradation, SEI layer growth, and lithium plating.

This implies that "hidden" variables capturing underlying degradation also need to be considered for an accurate prediction of RUL. In other words, different levels of each of these degradation mechanisms active within the cell, while not distinguishable from the normal SOH-C and SOH-R measurements at the end of the first life, can cause various trajectories in the second life. For instance, Li-plating can lead to accelerating degradation trajectories, while SEI growth can show self-limiting trends. Moreover, combinations of these types of nonlinear behaviors can influence the degradation paths for each cell. Therefore, to uniquely predict the RUL, the internal states of the degradation dynamics need to be known, which collectively constitute the estimation of deepSOH (Movahedi et al. 2024). Further improvement in RUL prediction can be made by considering extra measurements, such as the irreversible expansion of the cells (expansion due to aging that is not recoverable).

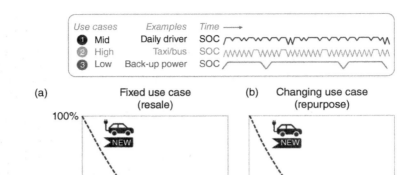

FIGURE 6.11 Large variability in battery second-life depending on use-case: Point ① in panels (a) and (b) depicts 70% state of health, which is the end of the battery's first life. Past that point, panel (a) shows the case where the BMS of an EV battery can extrapolate the estimated SOH to calculate the remaining useful life (RUL) of seven years if it continues to be driven along the same driving profile. However, panel (b) shows how RUL (increases) decreases if the duty cycle in the second life is (less) more demanding than the first. Point ② corresponds to a short second-life under a highly demanding taxi/bus drive cycle, whereas ③ corresponds to a longer second-life under a less demanding backup power use case.

Chemistry-Specific Challenges: LFP

Batteries containing LFP deserve special attention as they offer numerous benefits compared to NMC, including cost, safety, and decreased dependency on high-cost and high-conflict supply chains (Wheeler et al. 2024; Yi et al. 2024). As leading cell manufacturers have transitioned to higher energy densities and larger format prismatic LFP cells, the electrochemical and financial value of an individual cell has increased substantially. Therefore, there is a tremendous opportunity to improve the TCO through extended utilization and increased residual value in the second life with assured RUL estimation. However, regulatory requirements such as the EU Battery Directive and California Clean Cars II initiative are likely to require more sophisticated battery health monitoring for large format cells than mere extrapolations based on empirical degradation and safety models (Weng et al. 2023).

A major challenge associated with SOH estimation in LFP cells is the flatness of their open circuit voltage curve (except at very low or very high SOCs) and hysteresis between the charge and discharge curves, as illustrated in Figure 6.12. The insensitivity of the voltage output to SOC input even causes problems for online estimation of SOC, which is more straightforward and typically the first step toward estimating SOH. It has been suggested that additional measurements be utilized to gain more accurate SOC estimation results. Figueroa-Santos et al. (2020) proposed that using mechanical measurements, such as force exerted on the battery casing due to the

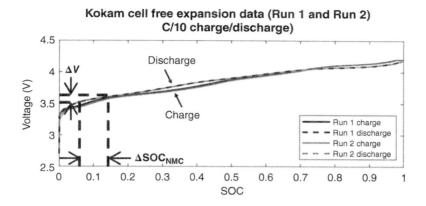
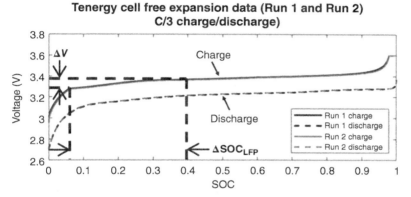

FIGURE 6.12 Voltage vs. state of charge (SOC) behavior of a Kokam NMC and Tenergy LFP cell for two runs of low-current charging/discharging. For the same change in voltage during charging represented by ΔV (chosen to be 100 mV only for the purpose of illustration), the *approximate* corresponding change in SOC for NMC (ΔSOC_{NMC}) is 0.06 or 6% while that for LFP (ΔSOC_{LFP}) is 0.4 or 40%. In other words, for the same change in voltage that can be induced by bias, drifts or noise in the sensor, change is SOC for LFP is more than six times that of NMC.

expansion of cells, can improve the performance of SOC estimation algorithms, which usually only include voltage as an input. However, specific measures must be implemented to model the mechanical force–SOC relationship to robustify the SOC and SOH estimation (Movahedi et al. 2020).

A critical factor in reporting the accuracy of SOH/SOC estimations is the inherent bias and drift in measuring voltage and current, which are outcomes of imperfections in sensors and data acquisition (Shi et al. 2023). The effects of sensor bias and drift are more pronounced in LFP cells due to the flatness of the OCV–SOC curve (refer to Figure 6.12) and some researchers have accounted for such disturbances in the measured current (Figueroa-Santos et al. 2020; Jöst et al. 2024). In many studies (Ko and Choi 2021; Mao et al. 2022), however, the pack-level current measurement is assumed to be completely accurate, similar to the laboratory grade sensor where their integration is used as the ground truth for SOC. In such cases, using the same perfect current signal as an input to the estimator is trivially wrong, and the estimator's reported accuracy must be treated with caution.

> ### Future Considerations
>
> Various stakeholders in the battery economy, such as manufacturers, service providers, and legislators, need to collaborate starting from the battery design phase to ensure that the impediments to battery reuse are removed (Kamath and Anctil 2024). Since the repurposing industry is still new, there is an urgent need to demonstrate successful pilot projects deploying repurposed batteries so that consumers' concerns around uncertainty in SLB performance, safety, and service life can be mitigated. This section will shed light on how retired batteries can be made more accessible, standardized, economically feasible, and easy to control and manage during operation.

Creating a Battery Reuse Network

The logistics of retired battery collection and transportation have still not been fully defined, especially in North America, which has led to a knowledge gap about the path that retired EV packs should trace. Slattery et al. (2021) conducted a literature review wherein 70% of the studies cited collecting and transporting EoL batteries as an impediment to reuse, and 63% emphasized the need for a policy that streamlines their collection and transportation.

Cascade utilization of EV batteries from the first to the second life can be promoted by creating a business model where EV manufacturers sell or lease the battery pack to EV owners, collect it when it reaches EoL, and then sell it to SLB providers or ESS manufacturers (Wu et al. 2020). A notable example of building a battery reuse network has been set by China's Beijing Automotive Group (BAIC), which closes the loop by collecting batteries with less than 80% capacity at its battery swapping stations and then using them (at the pack and/or module level) for energy storage and grid services (AO 2021b). With access to historical data on its packs, BAIC shortens the EoL SOH evaluation process to a mere 30 minutes (in contrast to the usual 10 hours) with 90% precision (AO 2021a).

Policies and Financial Incentives

To ensure sustainability in sourcing retired batteries and bolstering a circular economy, the European Union has mandated batteries with a capacity larger than 2 kWh (usually found in EVs and stationary ESS) to possess a Battery Passport starting 2026 (Regulation (EU) 2023/1542, 2023; Weng et al. 2023), which is an effective way of standardizing the battery industry. The Chinese government has also created a roadmap until 2030 to ensure battery reutilization at various echelons ("echelon utilization" refers to the detection, classification, and repair of SLBs for SL applications, Yu et al., 2020), which calls for comprehensive and accurate SOH/RUL monitoring for batteries being used in EVs, highly efficient and automated processes to sort and dismantle packs and standardize the utilization of battery modules (AO, 2022a). While China and the EU have published regulations focusing on SLB safety and health monitoring, the United States still does not have any federal battery reuse guidelines, barring a few initiatives in the state of California (Tankou 2023).

While such legislation is a positive step toward ensuring traceability and recycling of raw materials, reporting the key performance indicator of an SLB, that is, its RUL, is yet to be formalized. Legislation should also necessitate that battery management systems contain enough historical data to make an accurate prediction of RUL. If the battery passport is required to contain information regarding SOH that is accessible, it will reduce SLB providers' expenditure on performing SOH estimation. Moreover, policymakers should focus on strengthening safety checks and balances so that consumers' confidence in SLBs gets bolstered and they become more ubiquitous in applications such as home ESS (Murray 2023). They should also be careful about crafting regulations that mandate a certain percentage of recycled content in a newly manufactured battery (European Commission 2020), as requiring high amounts of recycled content can discourage repurposing and ultimately render it nonprofitable.

The SLB network can be strengthened using policies that facilitate the sharing of information between stakeholders about the battery's condition, thereby creating a domestic network that is less vulnerable to global supply-chain volatility (Slattery et al. 2024). The EU has proposed the rollout of extended producer responsibility (EPR), which holds producers responsible for collecting and managing their batteries at the end of life. While this is a step in the right direction, it needs to be augmented with a framework on how responsibilities will be transferred between the various stakeholders in a circular economy. For instance, whether the EPR will be transferred from the OEM to the SLB provider or the consumer must also be specified (Niese et al. 2020). This is particularly important given the concerns over SLBs causing fire hazards and the lack of clarity on liability in such situations (Reinhardt et al. 2019). China introduced an EPR system in 2016, outlining the roles of battery manufacturers in the design, sale, recovery, and echelon-reuse-and-recycling stages, most notably implementing a "product coding system" to encode all battery parts produced/imported such that there is a unique correspondence between the code and product, which can be archived and retrieved at the time of second use to facilitate traceability (AO 2022a).

PRACTICAL INSIGHTS | Traceability: The Evolution of a Policy and Practice

Tracking the origins of battery materials and their subsequent development – a process referred to as traceability – is critical to establishing their circularity. Reuse and repurposing, as well as safety, depend on knowing a battery's chemical makeup and the processes by which it has been manufactured and used. In the first half of the 2020s, several countries and economic blocks – most notably China and the EU – made formalized efforts to establish traceability within their battery supply chain for these, and other, purposes. China's battery materials tracking model, GB/T 34014-2017, is mandated and centrally managed by the Chinese government and establishes traceability via a QR code associated with a battery. The EU is utilizing a digital product passport (DPP) for batteries for traceability of materials. While the United States had no formal regulations in place in 2024, the 2021 Bipartisan Infrastructure Law (BIL) and 2022 Inflation Reduction Act (IRA) both supported tracking and tracing of materials, with the BIL providing economic incentives in the form of tax credits to companies sourcing and processing battery components in the United States. In 2024, the U.S. Department of Energy, Argonne National Laboratory, and a group of manufacturers and standards organizations began to explore more formalized means of establishing traceability through standardization. The state of California was also working on state Senate Bill 615, which, if passed, would make battery digital identifiers mandatory. A number of industry manufacturers, moreover, were already complying with the EU's DPP. At the time of this writing, traceability remained an active, if unsettled, area of practice and policy.

SLBs need to be included in the list of products eligible for tax incentives under laws such as the Inflation Reduction Act in the United States (Morris 2023). Governments also need to offer strong financial incentives that compel OEMs to participate in offering retired EV packs to SLB providers or create an internal ecosystem wherein their packs get repurposed into ESS units. Considering the fact that a vast majority of retired batteries are used in BESS engaging in grid services, utilities around the globe must step up to compensate BESS operators and consumers at competitive rates so that the business model remains sustainable and local governments are not burdened to provide subsidies in this domain (Kamath and Anctil 2024). The cost of collecting and transporting retired EV packs can be reduced by establishing repurposing facilities close to battery EoL collection hubs. Providers of SLBs can aim for automation of labor-intensive tasks such as disassembling a retired pack for sorting and testing modules.

Finally, policies must not remain mere recommendations but also enforce penalties for noncompliance in order to guarantee market participation (AO 2022a).

Case Study: B2U Storage Solutions

B2U Storage Solutions is a California-based energy storage developer that repurposes SLBs into large-scale BESS interconnected with Pacific Gas and electric's distribution system and sells electricity and services into the California Independent System Operator wholesale market. The packs in their original casing (in which they existed within an EV) are racked in an environmentally controlled enclosure that functions as an integrated EV Pack Storage (EPS) unit and in turn, the EPS functions as a building block within the overall ESS. The EPS is designed to facilitate easy installation, removal, and replacement of packs, which can be connected in series and parallel (depending on the voltage limit and desired capacity). Such rapid integration is facilitated by the fact that the used packs are often bought or leased directly from OEMs such as Honda Motor Co. and Nissan Motor Co. (Fine 2023), which gives B2U access to the BMS containing the battery's SOH (expected to be 5% accurate until EOL starting 2026, as per California Air Resources Board 2022b). The health of each pack is monitored via a battery pack controller (BPC), which interfaces with the pack's BMS. The BPC also performs active balancing between packs to ensure that they are charged and discharged in accordance with their individual capacities. Active balancing ensures that weaker batteries don't restrict the performance of stronger ones. The EPS can be utilized either in front of the meter (IFM) by connecting it to the grid or behind the meter (BTM) to fulfill a consumer's demand for energy. Ultimately, the EPS operates within a larger ESS, which consists of a PCS and supervisory control and data acquisition (SCADA) to supply power to the grid. EPS cabinets are assembled and tested before being shipped to a project site, where they are rapidly integrated into a functional ESS. The overarching EMS manages the cabinets by processing large volumes of data from thousands of battery packs to enable real-time SOH estimation that guides asset management and fault detection. The EPS ultimately leads to lower installation costs and superior levelized cost of storage compared to a BESS made of fresh batteries.

Words to Know

BESS Battery energy storage systems are devices composed of rechargeable batteries that store energy produced from renewable sources such as wind and solar and discharge it when the supply of renewable energy is scarce or absent. They help ensure a reliable supply of renewable energy at the grid level or backup power at the commercial/residential level.

BMS A battery management system is an electronic control circuit that monitors battery health and controls its charging and discharging. Its functions include monitoring voltage, current, temperature, capacity, and state of charge, among others. It prevents dangerous scenarios such as extreme fast charging and/or deep discharge and ensures that multiple cells within a pack remain balanced.

BoL Beginning of Life, which is the initial state of the battery prior to any usage and degradation, when it has maximum electrochemical capacity.

C-rate C-rate is the rate at which a battery is discharged with respect to its maximum capacity. For instance, a battery with a capacity of 5 Amp-hours discharging at 1C will expend 5 Amps of current in one hour before it gets completely discharged. The same battery at a rate of 2C will release 10 Amps of current in 30 minutes.

DeepSOH Deep state of health extends the conventional definition of state of health (SOH) to include states that represent degradation mechanisms, such as the thickness expansion due to the solid electrolyte interphase layer and thickness expansion due to lithium plating.

DOD Depth of discharge is the proportion of the battery's total capacity that has been discharged, expressed as a percentage of the total capacity.

eSOH Electrode state of health provides detailed information on the health of the battery by taking into account the electrodes' capacities and utilization windows (i.e. the electrical potential limits within which the electrode can operate).

EoL End of life is the point when a battery's capacity falls below a certain threshold. Typically for EV batteries, EoL is reached when capacity falls below 70–80% of the maximum rated value.

EoW End of warranty for an electric vehicle is the eight-year or 100,000-mile mark, whichever occurs first.

EPR Extended producer responsibility is a policy approach that makes producers take responsibility for managing their product throughout its lifecycle. It contributes toward the establishment of a circular economy through recycling, reusing, and incorporating design philosophies that minimize environmental burden.

ESS Energy storage systems store electricity generated from renewable sources, such as wind and solar, to be used later when energy is scarce or when it is economically advantageous to support the grid.

HPPC Hybrid pulse power characterization is a reference performance test procedure that is used to determine the pulse power and energy capability under no-load conditions as a function of aging for direct comparison with the targets in a gap analysis.

LAM$_{NE}$ Loss of active material of the negative electrode implies that the active mass of the negative electrode or anode has undergone particle cracking or is electrically isolated to which it can no longer host lithium ions when the battery is being charged. It can lead to a reduction in both available capacity and power.

LAM_{PE} Loss of active material at the positive electrode implies that the active mass of the positive electrode or cathode has undergone particle cracking or is electrically isolated to which it can no longer host lithium ions when the battery is being discharged. It can lead to a reduction in both available capacity and power.

LFP Lithium–iron–phosphate battery wherein the cathode is composed of iron phosphate.

LLI Loss of lithium inventory is the consumption of lithium ions in parasitic reactions within the battery, such as the growth of the solid electrolyte interphase layer and lithium plating. It means that lithium ions are no longer available for being cycled between the anode and cathode, which causes a reduction in the battery's capacity.

NMC Lithium–nickel–manganese–cobalt-oxide battery, wherein the cathode is composed of nickel, manganese, and cobalt.

NCX Cathodes contain lithium nickel cobalt oxides, where "X" can be manganese or aluminum.

OEM Original equipment manufacturer, which is a company that produces goods sold by other companies under their own brand name.

SEI Solid electrolyte interface is a passivation layer formed on the surface of the negative electrode because of electrolyte decomposition. It is essential for battery stability as it allows the transport of lithium ions, prevents further electrolyte decomposition, and ensures the continuation of electrochemical reactions within the battery.

SLB Second-life batteries refer to electric vehicle batteries that have reached the end of their warranty period but still have substantial useful capacity left in them. To make use of their remaining life, they are used in various applications such as energy storage, instead of being discarded or recycled.

SOH State of health is the ratio of a battery's capacity to its maximum rated capacity expressed as a percentage. It is a metric to quantify capacity retention at the cell or pack level and can be used to differentiate between fresh and aged cells.

RUL Remaining useful life can be defined in terms of the years, miles, or energy throughput (in kWh) until the battery's state of health drops below a minimum threshold.

Repurposing Repurposing refers to using a battery that has retired from being used in an electric vehicle for other less-demanding purposes such as energy storage.

Remanufacturing Remanufacturing refers to the process of restoring an electric vehicle's battery pack by performing repairs and/or reconditioning as required and then using it in the vehicle again.

Refurbishing Refurbishing refers to the process of restoring an electric vehicle's battery pack at the cell level by performing repairs and/or reconditioning and then using it in a new type of application.

RPT Reference performance test refers to periodic interruptions during calendar and cycle life aging to compute degradation in a battery. Degradation rates are determined by comparing results from the RPTs during cycle life testing with respect to the initial RPT performed at the beginning of life.

UBE Usable battery energy is the total direct current (DC) energy discharged by, measured in DC Wh, computed by a full depletion test or a combination of a partial depletion test and DC discharge test procedure, as defined by the SAE surface vehicle standards.

Further Reading

The National Renewable Energy Laboratory (NREL) aggregates research on battery second life from a vairety of sources and partnerships. **https://www.nrel.gov/transportation/battery-second-use.html**.

References

ACCURE Battery Intelligence (2024). Employing battery analytics for better insurance conditions. https://www.accure.net/battery-knowledge/employing-battery-analytics-for-better-insurance-conditions (accessed 24 December 2024).

Advanced Materials Manufacturing Technologies Office (2024). Re-X before recycling prize. **https://www.energy.gov/eere/ammto/re-x-recycling-prize** (accessed 19 August 2024).

Advanced Research Projects Agency – Energy (2024). U.S. Department of Energy Announces 30 Million to Develop Technologies to Enable Circular Electric Vehicle Battery Supply Chain. **https://arpa-e.energy.gov/news-and-media/press-releases/us-department-energy-announces-30-million-develop-technologies-enable** (accessed 24 December 2024).

Aguilar Lopez, F., Lauinger, D., Vuille, F., and Müller, D.B. (2024). On the potential of vehicle-to-grid and second-life batteries to provide energy and material security. *Nature Communications* 15 (1): 4179.

American Resources Corporation (2024). American Resources Corporation's ReElement Technologies Executes MOU for Initial Offtake of LFP Black Mass with Duesenfeld GmbH. **https://www.accesswire.com/823705/american-resources-corporations-reelement-technologies-executes-mou-for-initial-offtake-of-lfp-black-mass-with-duesenfeld-gmbh** (accessed 14 December 2024).

AO Stephanie (2021a). Sparking a second life of power battery, Part 1. An overview of battery reuse in China. **https://www.integralnewenergy.com/?p=31304** (accessed 12 August 2024).

AO Stephanie (2021b). Sparking a second life of power battery, Part 5. Business model and key players in China. **https://www.integralnewenergy.com/?p=31401** (accessed 12 August 2024).

AO Stephanie (2022a). Sparking a second life of power battery, Part 2. Political aspects of battery reuse in China. **https://www.integralnewenergy.com/?p=32365** (accessed 12 August 2024).

AO Stephanie (2022b). Sparking a second life of power battery, Part 3. Technical aspects of battery reuse in China. **https://www.integralnewenergy.com/?p=32375** (accessed 12 August 2024).

Attia, Peter M, Bills, Alexander, Planella, Ferran Brosa, Dechent Philipp, Dos Reis Goncalo, Dubarry Matthieu, Gasper Paul, Gilchrist Richard, Greenbank Samuel, Howey David, others. "Knees" in lithium-ion battery aging trajectories Journal of The Electrochemical Society. 2022. 169, 6. 060517.

August, B. (2023). What happens to old EV batteries? **https://www.recurrentauto.com/research/what-happens-to-old-ev-batteries** (accessed 25 July 2024).

B2U Energy (2024). B2U storage solutions. **https://www.b2uco.com/** (accessed 24 July 2024).

Bach, A., Onori, S., Reichelstein, S.J., and Zhuang, J. (2024). Fair market valuation of electric vehicle batteries in second life applications. Research Papers 4203, Stanford University, Mannheim.

Bai, B., Xiong, S., Song, B., and Xiaoming, M. (2019). Economic analysis of distributed solar photovoltaics with reused electric vehicle batteries as energy storage systems in China. *Renewable and Sustainable Energy Reviews* 109: 213–229.

Bashevkin, E. (2022). Why large-scale fire testing is needed for battery energy storage safety. **https://blog.fluenceenergy.com/battery-energy-storage-product-fire-safety-testing** (accessed 25 July 2024).

Baum, Z.J., Bird, R.E., Xiang, Y., and Jia, M. (2022). Lithium-ion battery recycling-overview of techniques and trends. *ACS Energy Letters* 7 (2): 712–719.

Birkl, C.R., Roberts, M.R., Euan, M.T. et al. (2017). Degradation diagnostics for lithium ion cells. *Journal of Power Sources* 341: 373–386.

Börner, M.F., Frieges, M.H., Späth, B. et al. (2022). Challenges of second-life concepts for retired electric vehicle batteries. *Cell Reports Physical Science* 3 (10): 101095.

Cabrera-Castillo, E., Niedermeier, F., and Jossen, A. (2016). Calculation of the state of safety (SOS) for lithium ion batteries. *Journal of Power Sources* 324: 509–520.

California Air Resources Board (2022a). Final regulation order, Section 1962.8, Title 13, California Code of Regulations. **https://ww2.arb.ca.gov/sites/default/files/barcu/regact/2022/accii/2acciifro1962.8.pdf** (accessed 14 December 2024).

California Air Resources Board (2022b). Proposed regulation order, Appendix A-8, Section 1962.7, Title 13, California Code of Regulations. **https://ww2.arb.ca.gov/sites/default/files/barcu/regact/2022/accii/appa8.pdf** (accessed 14 December 2024).

Chen, M., Ma, X., Chen, B. et al. (2019). Recycling end-of-life electric vehicle lithium-ion batteries. *Joule* 3 (11): 2622–2646.

Chen, Y., Kang, Y., Zhao, Y. et al. (2021). A review of lithium-ion battery safety concerns: the issues, strategies, and testing standards. *Journal of Energy Chemistry* 59: 83–99.

Christopherson, J.P. (2015). Battery test manual for electric vehicles. **https://inldigitallibrary.inl.gov/sites/sti/sti/6492291.pdf** (accessed 28 September 2024).

Colthorpe, A. (2019). China to 'dominate recycling and second life battery market worth USD 45bn by 2030'. **https://www.energy-storage.news/china-to-dominate-recycling-and-second-life-battery-market-worth-us45bn-by-2030/** (accessed 12 August 2024).

Cui, X., Ramyar, A., Mohtat, P. et al. (2022). Lite-sparse hierarchical partial power processing for second-use battery energy storage systems. *IEEE Access* 10: 90761–90777.

Dai, Q., Spangenberger, J., Ahmed, S. et al. (2019). *EverBatt: A Closed-Loop Battery Recycling Cost and Environmental Impacts Model*. Argonne National Laboratory.

Denholm, P., Cole, W. and Blair, N. (2023). Moving beyond 4-hour Li-ion batteries: challenges and opportunities for long (er)-Duration energy storage. Technical Report. **https://doi.org/10.2172/2000002**

Dong, Q., Liang, S., Li, J. et al. (2023). Cost, energy, and carbon footprint benefits of second-life electric vehicle battery use. *iScience* 26 (7): 107195.

Drallmeier, J.A., Wong, C., Solbrig, C.E. et al. (2022) Challenges of a fast diagnostic to inform screening of retired batteries. IFAC-PapersOnLine. 55, 24. 185–190. 10th IFAC Symposium on Advances in Automotive Control AAC 2022.

Dubarry, M., Truchot, C., and Liaw, B.Y. (2012). Synthesize battery degradation modes via a diagnostic and prognostic model. *Journal of Power Sources* 219: 204–216.

ECO STOR (2023). Safety first: what you need to know about Second Life batteries and their reliability. **https://www.eco-stor.com/en/resources/blog/safety-first-what-you-need-to-know-about-second-life-batteries-and-their-reliability** (accessed 25 July 2024).

Electric Vehicle Database (2024). Electric vehicle database. **https://ev-database.org/cheatsheet/range-electric-car** (accessed 19 September 2024).

Energy Storage News (2024). VIDEO: Insurance innovations that unlock the potential of battery storage systems. **https://www.energy-storage.news/video-insurance-innovations-unlock-potential-of-battery-storage-systems** (accessed 25 July 2024).

Environmental Protection Agency (2021). *Waste Management Hierarchy and Homeland Security Incidents*. United States Environmental Protection Agency.

European Commission (2020). Annexes to the Proposal for a Regulation of the European Parliament and of the Council Concerning Batteries and Waste Batteries, Repealing Directive 2006/66/EC and Amending Regulation (EU) No 2019/1020.

Fallah, N. and Fitzpatrick, C. (2023). Is shifting from Li-ion NMC to LFP in EVs beneficial for second-life storages in electricity markets? *Journal of Energy Storage* 68: 107740.

Fan, T., Liang, W., Guo, W. et al. (2023). Life cycle assessment of electric vehicles' lithium-ion batteries reused for energy storage. *Journal of Energy Storage* 71: 108126.

Federal Highway Administration (2022). *2022 National Household Travel Survey*. Washington, DC: U.S. Department of Transportation **https://nhts.ornl.gov**.

Figueroa-Santos, M.A., Siegel, J.B., and Stefanopoulou, A.G. (2020). Leveraging cell expansion sensing in state of charge estimation: practical considerations. *Energies* 13 (10): **https://doi.org/10.3390/en13102653**.

Fine, H. (2023). B2U repurposes used EV batteries. **https://labusinessjournal.com/featured/b2u-repurposes-used-ev-batteries/** (accessed 19 August 2024).

Gasper, P., Schiek, A., Smith, K. et al. (2022). Predicting battery capacity from impedance at varying temperature and state of charge using machine learning. *Cell Reports Physical Science* 3 (12): 101184.

Hall, F.S.and Stern, M.J. (2022). Energy storage system employing second-life electric vehicle batteries. III 29 2022. US Patent 11,289,921.

Hua, Y., Liu, X., Zhou, S. et al. (2021). Toward sustainable reuse of retired lithium-ion batteries from electric vehicles. *Resources, Conservation and Recycling* 168: 105249.

Huang, W., Feng, X., Han, X. et al. (2021). Questions and answers relating to lithium-ion battery safety issues. *Cell Reports Physical Science* 2 (1): 100285.

Idaho National Engineering Laboratory (1996). USABC electric vehicle battery test procedures manual revision 2. **https://digital.library.unt.edu/ark:/67531/metadc666152/m2/1/high_res_d/214312.pdf** (accessed 28 September 2024).

International Energy Agency (2024). Global EV Outlook: Trends in electric vehicle batteries. **https://www.iea.org/reports/global-ev-outlook-2024/trends-in-electric-vehicle-batteries** (accessed 16 December 2024).

Janaky (2024). Can EV batteries be reused? **https://www.azom.com/article.aspx?ArticleID=23810** (accessed 16 December 2024).

Jiao, N. (2018). China Tower can 'absorb' 2 million retired electric vehicle batteries. IDTechEx. **https://www.idtechex.com/de/research-article/china-tower-can-absorb-2-million-retired-electric-vehicle-batteries/15460** (24 December 2024).

Jöst, D., Palaniswamy, L.N., Quade, K.L., and Sauer, D.U. (2024). Towards robust state estimation for LFP batteries: model-in-the-loop analysis with hysteresis modelling and perspectives for other chemistries. *Journal of Energy Storage* 92: 112042.

Kamath, D. and Anctil, A. (2024). What's stopping electric vehicle batteries from having a second life? *Cell Reports Sustainability* 1 (4): 100080.

Kamath, D., Arsenault, R., Kim, H.C., and Anctil, A. (2020). Economic and environmental feasibility of second-life lithium-ion batteries as fast-charging energy storage. *Environmental Science & Technology* 54 (11): 6878–6887.

Karger, Alexander, Wildfeuer, Leo, ü Aygül, Deniz, Maheshwari, Arpit, Singer, Jan P, Jossen, Andreas. Modeling capacity fade of lithium-ion batteries during dynamic cycling considering path dependence. Journal of Energy Storage 2022. 52. 104718.

Kendall, A., Slattery, M. and Dunn, J.2024) End of life EV battery policy simulator: a dynamic systems, mixed-methods approach.

Ko, Y. and Choi, W. (2021). A new SOC estimation for LFP batteries: application in a 10 Ah Cell (HW 38120 L/S) as a hysteresis case study. *Electronics* 10: 6, 705.

Koroma, M.S., Costa, D., Philippot, M. et al. (2022). Life cycle assessment of battery electric vehicles: implications of future electricity mix and different battery end-of-life management. *Science of The Total Environment.* 831: 154859.

Kovachev, G., Ellersdorfer, C., Gstrein, G. et al. (2020). Safety assessment of electrically cycled cells at high temperatures under mechanical crush loads. *eTransportation*. 6: 100087.

Kumtepeli, Volkan, Hesse, Holger, Morstyn, Thomas, Nosratabadi, Seyyed Mostafa, Aunedi, Marko, Howey, David A. Depreciation cost is a poor proxy for revenue lost to aging in grid storage optimization. arXiv preprint arXiv:2403.10617. 2024.

Lander, L., Cleaver, T., Rajaeifar, M.A. et al. (2021). Financial viability of electric vehicle lithium-ion battery recycling. *Iscience* 24 (7): 102787.

Lee, S., Siegel, J.B., Stefanopoulou, A.G. et al. (2020). Electrode state of health estimation for lithium ion batteries considering half-cell potential change due to aging. *Journal of The Electrochemical Society* 167 (9): 090531.

Lorscheid, V. (2024). Unlocking the potential of second-life lithium-ion batteries: challenges and solutions. **https://www.accure.net/battery-knowledge/second-life-lithium-ion-batteries** (accessed 25 July 2024).

Madalin (2023). Second-life electric vehicle batteries 2023–2033. **https://hafenstrom.com/second-life-electric-vehicle-batteries-2023-2033/** (accessed 12 August 2024).

Mao, S., Han, M., Han, X. et al. (2022). An electrical–thermal coupling model with artificial intelligence for state of charge and residual available energy co-estimation of $LiFePO_4$ battery system under various temperatures. *Batteries* 8 (10).

Marjolin, Aude (2023). Lithium-ion battery capacity to grow steadily to 2030. **https://www.spglobal.com/marketintelligence/en/news-insights/research/lithium-ion-battery-capacity-to-grow-steadily-to-2030** (accessed 25 July 2024).

Meegoda, J., Charbel, G., and Watts, D. (2024). Second life of used lithium-ion batteries from electric vehicles in the USA. *Environments* 11 (5).

Morris, C. (2023). Packaging second-life EV batteries into a plug-and-play energy storage system. **https://chargedevs.com/features/packaging-second-life-ev-batteries-into-a-plug-and-play-energy-storage-system/** (accessed 25 July 2024).

Movahedi, H., Figueroa-Santos, M.A., Siegel, J.B. et al. (2020). Hybrid nonlinear observer for battery state-of-charge estimation using nonmonotonic force measurements. *Advanced Control for Applications: Engineering and Industrial Systems.* 2 (3): e38.

Movahedi, Hamidreza, Weng, Andrew, Pannala, Sravan, Siegel, Jason B., Stefanopoulou, Anna G. (2024). The case for DeepSOH: addressing path dependency for remaining useful life. **https://arxiv.org/abs/2405.12028** (accessed 25 July 2024).

Murray, Cameron (2023). Opposing views emerge on safety of second life EV batteries for home energy storage – UK study. **https://www.energy-storage.news/opposing-views-emerge-on-safety-of-second-life-ev-batteries-for-home-energy-storage-uk-study/** (accessed 25 July 2024).

National Instruments (2024). Battery recycling and second-life test solutions. **https://www.ni.com/en/solutions/transportation/electric-vehicle-test/battery-recycling-second-life-test.html** (accessed 25 July 2024).

National Renewable Energy Laboratory (2020). Battery second-use repurposing cost calculator. **https://www.nrel.gov/transportation/b2u-calculator.html** (accessed 12 August 2024).

Neubauer, J. and Pesaran, A. (2011). The ability of battery second use strategies to impact plug-in electric vehicle prices and serve utility energy storage applications. *Journal of Power Sources* 196 (23): 10351–10358.

Nichols, Conrad (2023). Considerations for Remanufacturing Second-Life EV Batteries. IDTechEx. **https://www.idtechex.com/en/research-article/considerations-for-remanufacturing-second-life-ev-batteries/29134** (accessed 24 December, 2024).

Niese, Nathan, Pieper, Cornelius, Arora, Aakash, Xie, Alex. (2020). The case for a circular economy in electric vehicle batteries. **https://www.bcg.com/publications/2020/case-for-circular- economy-in-electric-vehicle-batteries** (accessed 25 July 2024).

Oak Ridge National Laboratory (2021). Transportation energy data book: Edition 38. **https://tedb.ornl.gov/data/** (accessed 16 December 2024).

Pannala, S., Movahedi, H., Garrick, T.R. et al. (2024). Consistently tuned battery lifetime predictive model of capacity loss, resistance increase, and irreversible thickness growth. *Journal of The Electrochemical Society* 171 (1): 010532.

Patel, A.N., Lander, L., Ahuja, J. et al. (2024). Lithium-ion battery second life: pathways, challenges and outlook. *Frontiers in Chemistry* 12: 1358417.

Plett, Gregory L. Battery Management Systems, Volume I: Battery Modeling. 2015.

Prasad, G.K. and Rahn, C.D. (2013). Model based identification of aging parameters in lithium ion batteries. *Journal of power sources.* 232: 79–85.

Preger, Yuliya (2024). Impact of aging on the safety of lithium-ion batteries. https://www.sandia.gov/app/uploads/sites/82/2024/08/PR2024_705_Preger_Yuliya_Safety-Reliability-2.pdf (accessed 25 August 2024).

Preger, Y., Torres-Castro, L., Rauhala, T., and Jeevarajan, J. (2022). Perspective—on the safety of aged lithium-ion batteries. *Journal of the Electrochemical Society* 169 (3): 030507.

Regulation (EU) 2023/1542 of the European Parliament and of the Council of 12 July 2023 (2023). Concerning batteries and waste batteries, amending Directive 2008/98/EC and Regulation (EU) 2019/1020 and repealing Directive 2006/66/EC (Text with EEA relevance). 2023. **https://eur-lex.europa.eu/eli/reg/2023/1542/oj** (accessed 25 July 2024).

Reinhardt, R., Christodoulou, I., Gass'o-Domingo, S., and García, B.A. (2019). Towards sustainable business models for electric vehicle battery second use: a critical review. *Journal of Environmental Management.* 245: 432–446.

ReJoule Energy (2021). The obstacle course on the path to repurposing used electric vehicle batteries (EVB). Part IV: impact of cell imbalance. **https://rejouleenergy.com/blog/the-obstacle-course-on-the-path-to-repurposing-used-electric-vehicle-batteries-evb-part-iv-impact-of-cell-imbalance** (accessed 19 August 2024).

ReJoule Energy (2024a). Case study: rapid state-of-health (SOH) estimation of decommissioned ev batteries using ReJoule's BattScan050M. **https://rejouleenergy.com/case-study** (accessed 19 August 2024).

ReJoule Energy (2024b). World-first battery health test via charging port. **https://rejouleenergy.com/beta** (accessed 19 August 2024).

ReJoule Energy (2024c). The science behind our technology. **https://rejouleenergy.com/the-science-behind-rejoule** (accessed 19 August 2024).

Reniers, J.M. and Howey, D.A. (2023). Digital twin of a MWh-scale grid battery system for efficiency and degradation analysis. *Applied Energy* 336: 120774.

Roy, A., Movahedi, H., Siegel, J.B., and Stefanopoulou, A.G. (2023). Empirical modeling of degradation in lithium-ion batteries and validation in complex scenarios. *IFAC-PapersOnLine* 56 (3): 457–462.

SAE J1634 (2021). Battery electric vehicle energy consumption and range test procedure. J1634_202104. **https://www.sae.org/standards/content/J1634_202104** (accessed 22 February 2025).

Salza, P., Takahito, K., Mhani, L., Ye, X. (2021). 2nd Life Batteries – A white paper from Storage Technological Community. **https://www.globalelectricity.org/wp- content/uploads/2022/09/GSEP SecondLifeBatteries.pdf** (accessed 25 July 2024).

Severson, K.A., Attia, P.M., Jin, N. et al. (2019). Data-driven prediction of battery cycle life before capacity degradation. Nature. *Energy* 4 (5): 383–391.

Shen, J. (2021). China to ban large energy storage plants from using retired EV batteries. **https://technode.com/2021/06/24/china-to-ban-large-energy-storage-plants-from-using-retired-ev-batteries/** (accessed 30 December 2024).

Shi, J., Kato, D., Jiang, S. et al. (2023). Robust estimation of state of charge in lithium iron phosphate cells enabled by online parameter estimation and deep neural networks. *IFAC-PapersOnLine* 56 (3): 127–132.

Slattery, M., Dunn, J., and Kendall, A. (2021). Transportation of electric vehicle lithium-ion batteries at end-of-life: a literature review. *Resources, Conservation and Recycling* 174: 105755.

Slattery, M., Dunn, J., and Kendall, A. (2024). Charting the electric vehicle battery reuse and recycling network in North America. *Waste Management* 174: 76–87.

Smartville (2024). Unlocking the true value of EV batteries. **https://smartville.io/** (accessed 25 July 2024).

Srinivasan, L., Shaw, S. and Billaut, E. (2024). Insights from EPRI's battery energy storage systems (BESS) failure incident database: analysis of failure root cause.

Strange, C., Li, S., Gilchrist, R., and Reis, G.D. (2021). Elbows of internal resistance rise curves in Li-ion cells. *Energies* 14: 4. **https://doi.org/10.3390/en14041206**.

Sun, S.I., Chipperfield, A.J., Kiaee, M., and Wills, R.G.A. (2018). Effects of market dynamics on the time-evolving price of second-life electric vehicle batteries. *Journal of Energy Storage* 19: 41–51.

Tankou, A., Georg, B., and Dale, H. (2023). Scaling up reuse and recycling of electric vehicle batteries: Assessing challenges and policy approaches. *Proceedings of ICCT* 1–138.

U.S. Department of Energy (2024). Lithium-ion battery storage technical specifications. **https://www.energy.gov/sites/default/files/2024-07/bess-technical-specifications-2024.docx** (accessed 25 July 2024).

U.S. Energy Information Administration (2023). Use of energy explained: Energy use in homes. **https://www.eia.gov/energyexplained/use-of-energy/electricity-use-in-homes.php** (accessed 25 July 2024).

UN GTR No. 22 (2022). In-vehicle battery durability for electrified vehicles. **https://unece.org/transport/documents/2022/04/standards/un-gtr-no22-vehicle-battery-durability-electrified-vehicles** (accessed 16 December 2024)

Underwriters Laboratory (2019). Second-life electric vehicle battery repurposing facility certification. **https://www.ul.com/services/second-life-electric-vehicle-battery-repurposing-facility-certification** (accessed 25 July 2024).

Wakihara, M. and Yamamoto, O. (2008). *Lithium Ion Batteries: Fundamentals and Performance*. Wiley.

Wang, M., Liu, K., Dutta, S. et al. (2022). Recycling of lithium iron phosphate batteries: status, technologies, challenges, and prospects. *Renewable and Sustainable Energy Reviews* 163: 112515.

Weng, A., Mohtat, P., Attia, P.M. et al. (2021). Predicting the impact of formation protocols on battery lifetime immediately after manufacturing. *Joule* 5 (11): 2971–2992.

Weng, A., Dufek, E., and Stefanopoulou, A. (2023). Battery passports for promoting electric vehicle resale and repurposing. *Joule* 7 (5): 837–842.

Weng, A., Movahedi, H., Wong, C. et al. (2024). Current imbalance in dissimilar parallel-connected batteries and the fate of degradation convergence. *The Journal of Dynamic Systems, Measurement, and Control* 146: 011106.

Wheeler, W., Venet, P., Bultel, Y. et al. (2024). Aging in first and second life of G/LFP 18650 cells: diagnosis and evolution of the state of health of the cell and the negative electrode under cycling. *Batteries* 10 (4): 137.

Whitlock, R. (2024). Altilium announces advances in recycling LFP and NMC electric vehicle batteries. **https://www.renewableenergymagazine.com/electric_hybrid_vehicles/altilium-announces-technological-advances-in-recycling-lfp-20240510** (accessed 25 July 2024).

Wu, W., Lin, B., Xie, C. et al. (2020). Does energy storage provide a profitable second life for electric vehicle batteries? *Energy Economics* 92: 105010.

Xu, C., Behrens, P., Gasper, P. et al. (2023). Electric vehicle batteries alone could satisfy short-term grid storage demand by as early as 2030. *Nature Communications* 14 (1): 119.

Yi, Baozhao, Du, Xinhao, Zhang, Jiawei, Wu, Xiaogang, Hu, Qiuhao, Jiang, Weiran, Hu, Xiaosong, Song, Ziyou. Bias-compensated state of charge and state of health joint estimation for lithium iron phosphate batteries. arXiv preprint arXiv:2401.08136. 2024.

Yu, H., Dai, H., Tian, G. et al. (2020). Big-data-based power battery recycling for new energy vehicles: information sharing platform and intelligent transportation optimization. *IEEE Access* 8: 99605–99623.

Zhang, Y., Tang, Q., Zhang, Y. et al. (2020). Identifying degradation patterns of lithium ion batteries from impedance spectroscopy using machine learning. *Nature Communications* 11 (1): 1706.

Zhu, J., Mathews, I., Ren, D., and Li, W. (2021). End-of-life or second-life options for retired electric vehicle batteries. *Cell Reports Physical Science* 2 (8): 100537.

Zhu, L., Yao, X., Su, B. et al. (2024). Reusing vehicle batteries can power rural China while contributing to multiple SDGs. *Cell Reports Sustainability* 1 (4): 100060.

CHAPTER 7

Designing New Engineering Teams and Practices

John Warner
Holly, Michigan, USA

Introduction

The engineering team forms the core of any product-based organization. Its members design, develop, prototype, test, validate, and often integrate one or more products for use either as a stand-alone device or as part of another product. This is the process that turns an idea or concept into a salable product, or as a good friend of mine likes to say "turning technology into cash." This chapter will focus on *how* to structure your team to achieve this goal based on the age and type of organization because an organization will have different needs at different points in its development.

What You Will Learn in This Chapter

This chapter introduces and examines the challenges and opportunities of planning, designing, and building an engineering team *based on the stage of your business*. This chapter will examine some of the key organizational structures that are often used in building companies and engineering teams. There are other resources available to describe how to build an engineering organization, but this chapter differs in that it focuses on looking at the design of the engineering organization at different stages of company development. The early-stage start-up company will have a very different engineering team than a 5-year-old company, or a 100-year-old company. Aligning the size, goals, structure, and communication methods of the team with the size of the organization will enable you to grow your team with the company.

Electric Vehicle Batteries: From Sourcing to Second Life and Recycling, First Edition.
Edited by Bob Galyen and Frank Menchaca.
© 2025 John Wiley & Sons, Inc. Published 2025 by John Wiley & Sons, Inc.

Through the course of this chapter, the reader will learn to:

- Understand the different stages of company development.
- Understand the different types of engineering organizations needed in each stage.
- Design and plan the creation and building of an engineering team.
- Understand the importance of integration and communications.
- Understand the differences in customer and product requirements and the challenges in translating them into product requirements.

The Essentials of the Team

Battery engineering follows the same basic foundational engineering processes as with any other technological development. This begins with defining the requirements and follows through to the detailed component and system design and validation of the product and performance against those requirements. The engineering team will follow the product through the process right to production and after. However, the way that you organize and build your engineering team will depend on what type of product you are developing and the age of your company. A company will evolve through five stages of development and maturity, beginning as a start-up, then moving into an expansion phase. This is followed by the growth stage and culminates in the maturity and decline stages. An engineering team will look different in each of these stages.

Designing New Engineering Organizations

The explosive growth in the advanced battery industry over the past decade has meant that there is an influx of new engineers, mid-career engineers, and engineers experienced in different fields that may have little understanding of chemistry or energy storage technologies. This raises several questions including, how do I bring these engineers up to speed as quickly as possible? What assignments can I give them to help grow their skills and capabilities? How do I effectively cross-train them in different disciplines that are necessary for battery development? And most importantly, how do I create a structure that supports the product growth and grows the team, while meeting deadlines?

The other question of key importance is how do I structure my team? There are many different types of organizational structures that have been introduced over the years under the study of organizational design. These structures range from the traditional hierarchical to the pure team-based structure and the organism-based structure. But it is not enough to simply look at the type of structure a team will be built around, we must also understand the stage of development of the company and the products you are developing. The way you organize your team will depend on whether your organization is focused on fundamental materials research and development (R&D) or product development for near term products. In comparing the R&D and the development organization, the major differences are in the end-product and the timing of its introduction. A research organization often tends to look at products that are 5–10 years out, whereas a product design organization is looking at products that are 1–4 years out. For the purposes of this discussion, we will focus on the engineering development organization, but these topics are applicable to both types of organizations.

The next section of this chapter will introduce the different stages of company development and provide some examples of companies that are in those stages. It will then introduce some of the key concepts around organizational design, including organizational types and designs. From this point, the chapter will expand on the five major stages of the company life cycle and describe engineering organizations in each stage. Next, the chapter will cover some of the major challenges associated with building an organization and provide some real-world case studies.

Organizational Development Stages

It is important to understand the nature of your organization when evaluating the structure of your engineering team. Organizations, much like humans, have life cycles – they grow, they mature, and eventually they die (Inc. 2020). An organization will make the move from one stage to the next if the external pressures and fit of the ". . . organization and its environment is so inadequate that either the organization's efficiency and/or effectiveness is seriously impaired or the organization's survival is threatened" (Inc. 2020). Companies evolve through several key stages of organizational growth as shown in Figure 7.1. A company will begin in the start-up phase and then move to expansion, followed by growth, maturity, and, finally, decline or restructure (G&A Partners 2021). Flamholtz and Randle (2016) describe a similar seven-stage organizational model of growth that begins with the entrepreneurial new venture creation, to expansion, professionalization, consolidation, diversification, integration, and decline and revitalization stages. Figure 7.1 overlays these two company growth stages with the traditional company growth curve and adds some description of the changes in the engineering team during each phase. Keep in mind that it may not always be clear which stage your company is operating in and, in some cases, companies may live in a stage for a long time or may return to a previous stage based on the performance of the company.

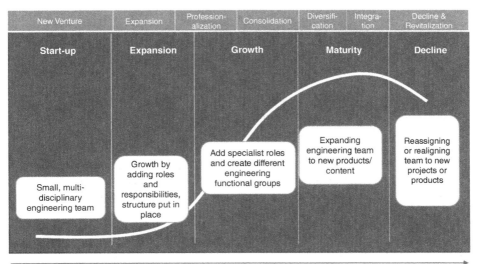

FIGURE 7.1 Stage of organizational and engineering team growth.

Start-ups, entrepreneurial, or new venture companies are in the earliest stages of development. They are entrepreneurial ventures typically started by one or more founders that have a vision of identifying and meeting a specific market demand through creation of a unique product, service, or platform (McGowan 2022; Flamholtz and Randle 2016). In this organization, there are relatively few people, but they tend to have greater levels of expertise because they are often inventing the product or service. The founder is focused on raising capital to fund the company so there is a push for rapid development and to show the stakeholders a return on their investment. In many cases, the founder will also be the inventor of the technology and may serve in the chief technology officer (CTO) role. This stage is marked by little revenue generation and high capital burn. Getting products in the hands of potential consumers quickly to prove success to the early investors is the focus of this stage.

The start-up engineering team is extremely lean with a small group of engineers wearing many hats. There is little formalization of roles, duties, and responsibilities during the start-up stage, people do what needs to get done. There is a high level of uncertainty and a high rate of change during this period. This is the riskiest stage of a business and is where most new businesses fail. The engineering activities may be led by the founder, Chief Executive Officer (CEO), CTO, or chief scientist role. Engineers during this stage of the company must be knowledgeable in many different aspects of the technology and may be responsible for more than one engineering function, mechanical, electrical, and thermal, which may be done by a single person for instance, which means that they may need to think more like a systems engineer. The systems engineer has always been one of the most critical roles in an organization as they are individuals that have a solid understanding of each of the technical fields associated with the product and can see how they interact. In larger organizations, these may be referred to as chief engineers while in others they may hold the title of lead pack engineer. These early-stage engineers tend to have fewer processes and procedures in place as they are totally focused on product invention, development, and speed to market. Requirements documentation is likely poor to nonexistent as the team pushes hard to create the first-generation product. In the case of a new technology, the customer or market requirements likely exist only in the head of the founder which often creates a challenge for the engineering team if the founder is not also leading the engineering efforts. Structurally, these engineers will report directly to whoever is in the lead engineering role – CEO, CTO, chief engineer, chief scientist, vice president of engineering, etc.

Case Study: Twig Power: Building an Agile Workforce

Twig Power is a tier three battery technology supplier based in southeast Michigan. Founded and helmed by a veteran engineer, Jesse Beeker, Twig has approached engineering team creation in an agile manner. "Most of the team's work flows through me," he says. "We supplement that with engineers who are not necessarily your classic, single-domain experts. Although we do electronics product design, for example, we've got a mechanical lead engineer and then we've got software and application engineers. That's our core structure for engineering." He continued: "From there, the production team is a separate entity all in themselves, although the engineering team does participate in the manufacturing."

Manufacturing is a *sine qua* non, something without which nothing else succeeds. "One requirement is for engineers to have an understanding of our manufacturing processes and capabilities. No matter what kind of project – if it's in batteries, for example – they understand the limitations of our assembly capabilities and thus the design for manufacturability is very high."

Twig's recruitment and talent acquisition approach leans heavily on developing people through hands-on experience. "I believe in a strong talent pipeline," Jesse says, "especially given the fact that the degreed candidates coming out of universities do not line up very well with, we'll call it, that hybrid engineering that is required in this space. Instead, we've heavily leveraged internships over the years. We're fortunate to be in a talent-rich educational community here in southeast Michigan. We've been able to leverage high school students who are capable of doing incredible things in electrical and mechanical engineering. This means we have entry-level engineers, engineers who work seasonally or part time. This helps manage costs."

Twig and its employees benefit from this. Of the high school interns he takes on, Jesse comments: "They learn how to write code in class. With us, they actually deploy it in an automotive based operating system. How many other students can claim that experience?" This makes for mature employees who, at a young age, can demonstrate real-world experience. This sets them up for success at Twig as well as at other companies. "In a world where leaders of the large automotive companies stand up and say, 'We don't have enough X engineers,' we can say we've created our own experts."

Organizations during this stage may use different product development and iteration strategies. For example, software companies tend to use a "minimal viable product" approach to product development where they may launch a software product quickly knowing that there are still some bugs to work out and then iterate the software frequently. In the battery arena, we need to think about it slightly differently, there is a minimum level of safety needed to launch any product, even early prototypes. I view battery safety as being like Herzberg's motivation theory around "hygiene factors" (1959). Having a safe battery is the minimum baseline requirement. By itself it does not necessarily cause the buyer to buy, but when paired with performance, life, cost, and environment features, it creates a reason to buy. These characteristics, in this order, were first applied to the battery industry by Bob Galyen and Jonas Beriesa through a series of discussions in the early 2000s and ultimately were defined as the five challenges of electrification: safety, performance, life, cost, and environment.

As the organization grows and the product matures and moves from development to launch, the organization also moves into the next stage of development – expansion. In the expansion stage (Flamholtz and Randle 2016), the company begins to increase and formalize the roles and responsibilities of the engineering team. The first-generation product moves into production and sales growth begins. The engineering team in this organization begins transforming to slightly more specialized roles and requires a greater level of integration to ensure alignment. The highly skilled multifunctional engineers from the start-up stage begin to move into leadership roles and the new engineers begin reporting into this team. This stage is marked by the engineering scope changing from fast-to-market product introduction, to looking at re-engineering the first-generation product to reduce costs, improve manufacturability, quality and reliability. This is driven by the need to begin becoming a profitable company and bringing the company to the break-even point. We see the emergence of new tools and systems into the engineering team to help manage the change and growth of the organization and company. The amount of uncertainty and change begins to moderate during this stage of development but there is often still some amount of overall uncertainty for the future as the company may still be funded by venture capital or early-stage investors. The company is often looking at bringing in new and/or strategic investors during this stage to help fund the growth and production of the products.

American Battery Solutions (ABS) is an example of a company in the expansion stage. Formed in 2019 by Subhash Dhar, the company pulled together some battery and automotive industry veterans with the goal of meeting the needs of what they called an "underserved" market segment – those lower volume battery customers who struggled to get the attention of the big cell makers due to the small size of their projects. Between 2019 and 2024, ABS developed several off-the-shelf and several custom products that launched in the market 2024. In 2021, they created and then in 2023 spun off a division that was focused on the stationary energy market and in 2023 were acquired by the leading global mining and construction company Komatsu America Corp. The engineering team followed the growth model described above, beginning in 2019 with a small team of skilled engineers and several industry veterans they began developing products for a specific segment of the market. At the same time, they engaged with several development projects that stretched the engineering resources thin. Eventually, an experienced engineering leader was brought in to build the team to meet the needs of the expansion phase with the company products expanding into new mining and construction markets while still growing and expanding their commercial vehicle and industrial battery business to new customers, markets, and geographies.

Moving to the growth stage of an organization, as sales growth begins to slow, a next generation of products are needed to continue to grow revenue and market share and new products are conceived and launched, while the products also become more specialized and may differentiate from one region or customer to the next. Growth into new markets and regions occurs in this stage and may require product differentiation and specialization to meet the needs of the regions or markets. Product and system engineering continues to evolve, and specialization continues to increase. During this stage, integration is a major focus at both the product and organizational levels. As engineering specialization increases, it becomes more important to ensure that the different functions are working toward the same goal. Greater specialization, by definition, drives greater separation, so ensuring that everyone knows where their parts fit in the greater whole is imperative. Maintaining strong communication is crucial during the growth stage. During this stage, the engineering team may begin to move away from the more hierarchical, top-down structure and move into a project or product-focused structure engaging more cross-functional engineering activities. The organization may move to either a matrix or a project-based structure depending on the company and products.

During this stage of development, the processes developed in the earlier stages get more sophisticated. While the corporate processes in the earlier stages were being created and implemented, processes used by companies in the expansion stage are focused on diversification and market expansion. More formalization is needed as companies cross-national and geographic boundaries to keep aligned with the corporate goals and directives. One example of this is seen in the addition of corporate compliance officers and engineering compliance engineers. At the corporate level, compliance is focused on ensuring processes align with regional legal requirements, monitoring and reporting, training and education, and alignment with regulatory bodies. At the engineering team level, the compliance engineer will be responsible for evaluating changing regional regulations and ensuring that the company's products meet those requirements. One real-world example of this is happening right now with the 2023 adoption of the European Union's updated Battery Directive regulation (EU) 2023/1542. This regulation adds several new requirements to the battery maker that are different than regulations in the United States that require U.S. engineers to compare their technologies specifications and certifications with the new regulation and make appropriate adjustments if they want to sell their products in Europe.

An example of a company in this stage may be Tesla. Tesla has successfully transitioned from the start-up and into the growth stage; they are moving into the expansion stage as they begin to look to Europe and Asia for their next growth opportunities. Their product development continues to excite and sometimes mystify the market and consumers. They have met growth projections and have grown to become the most valuable automobile company in the world. Because its roots are *not* 100 years old, Tesla focuses on innovative and unique solutions to the market needs. At the same time, they are working hard to evolve their core products to make them more reliable and reduce costs to make them more affordable in the mass market (Furr and Dyer 2020). During this expansion stage, their team has been supplemented through several acquisitions which bring new thinking and technologies into the company, but which may make communication more difficult depending on how the integrations are executed.

The maturity stage is marked by low levels of uncertainty and change. At this stage of development, managing and running the organization becomes very stable with little change. While uncertainty and change are at an all-time low during this stage, it could be the most dangerous stage because innovation and new product development may no longer be important to keeping the corporate engine running. The organization is focused on understanding whether they should change and if so how to change. This applies to the product, the structure, the markets, and the overall goals and vision of the company. The engineering team continues to focus on specialization during this stage with engineering functions moving into clear and distinct silos. Engineering leadership is trying to maintain some level of flexibility while being as efficient as possible. However, the structure of the engineering team becomes more complex as the organization continues to grow. Engineering processes and procedures become more engrained in the culture of the team and if not actively managed may begin to become bureaucratic in nature.

Many of the traditional automakers may find themselves in this stage. Their traditional markets and customers are beginning to see slight declines as new customers slowly begin to shift toward more electric options. We have seen nearly every major auto company announce major electrification plans, but they will all face significant internal challenges and headwinds as they fight through the "this is not the way we have always done it" way of thinking. Many of the original equipment manufacturers (OEMs) have been through massive restructuring of their organizations to attempt to maintain market share and a position as they develop new electric vehicle (EV) products, while still maintaining their current product portfolios. The engineering teams in a mature organization may be large with a high level of specialization which creates silos. The size of the organization also means that there is much more delegation occurring as the leader can no longer interact with everyone on the team. The size of the organization may begin to put stress on the processes, procedures, and communication channels that were developed in the last stage (Petrone 2016). Engineering leaders should focus on communication and the development of their management team to assist them in managing the organization.

The final company stage of development is decline and rebirth. Much like the parable of the frog in a boiling pot of water, a company may not realize they are here until it is too late (Senge 1994).

PRACTICAL INSIGHTS | The Parable of the Boiled Frog

In 1994, Peter Senge introduced the ancient parable of the boiled frog to the business world to describe what he referred to as "maladaptation to gradually building threats" (Senge 1994). The parable goes like this: If you take a frog and drop it in a pot of boiling water, it will immediately attempt to jump out and escape. However, if you put the frog into a pot of room temperature water and slowly increase the temperature to boiling, the frog will remain in the water and will not attempt to escape. As the temperature increases, the frog will get groggy but will still not attempt to escape. Eventually, as the water gets hotter the frog will lose the ability to escape and will sit there until it dies.

Senge writes that this happens "Because the frog's internal apparatus for sensing threats to survival is geared to sudden changes in his environment, not to slow, gradual changes" (1994, p. 22). The moral of the story is that people will react to sudden changes in their environment, but will accept small, slow changes until it is too late to respond. He goes on to say "learning to see slow, gradual processes requires slowing down our frenetic page and paying attention to the subtle as well as the dramatic" (1994, p. 23).

Today, we can use this parable to explain the growth of the Chinese battery industry. While the Chinese made long-term, strategic investments in the development and manufacturing of lithium-ion batteries (LIB), the rest of the world largely ignored the technology and remained focused on growing the internal combustion engine (ICE) markets. The Chinese investments grew and expanded, while the rest of the world made only small, incremental moves into the technology. Today, more than 80% of all LIBs and their materials are built in China while the governments of the rest of the world have now realized their water is beginning to boil and try to escape.

At this stage, there is often little innovation in the product and company and the products you started the company to produce may be in decline or may be replaced in the market. Blackberry is a great example of a company that had a great product and quickly grew market share but was ultimately replaced in the market by the much more adaptable and flexible product the iPhone. At this stage, you may find that, if the management recognizes the stage they are in, they create several new engineering teams or projects to investigate developing new products. These teams may get to experience the excitement of the start-up, but the rest of the engineering team may be shrinking during this stage of the company. During this stage and even in the maturity stage, there may be a loss of leaders and expertise as some of the more experienced employees begin looking to do more innovative work and move into an organization with less red tape. The stage ends either in the death of the organization or if a new technology or product is developed in the rebirth of the organization. The organization may go through reductions, downsizing, or even acquisition by other market players during this stage. Optimizing the engineering organization to support products and more importantly to invent new products is the focus of this stage.

A great example of an industry in this stage is the camera and film industry. While the camera and film industry grew and matured throughout the 20th century, it was right at the turn of the decade that the smartphone was introduced and changed the entire market for film and cameras. The camera and film market took a steady dive as the use of smartphones for taking pictures increased. Beginning in 1999, there was growth in the market with the introduction of the digital camera. This market continued to grow until the 2008–2010 period when the market took a massive decline due to the great improvements in smartphone cameras and the introduction of the iPhone 7. By 2020, both the film and digital camera markets had nearly disappeared (Smith 2021). This drove major changes to companies like Fujifilm, a company that nearly entirely focused its product development efforts on this market. But Fujifilm did something some of their competitors did not, they diversified into pharmaceutical and cosmetic products. This move made the camera market just a small portion of their business but required a major organizational change (Roesch 2020). It is possible to redirect a company that has entered the decline stage as this example shows, but it is often extremely difficult for the management to accept that they are here.

Types of Organizations

There are many possible types of organizational structures that can be applied when designing and building an engineering organization. Designing an engineering team in the battery industry is done with many of the same concerns and issues as in any technical engineering organization. As we begin to understand this process, let us look at the purpose of organizational structure. Organizational structure defines *how we arrange the company to do business*. It defines the communication flow, the way we delegate, our decision-making processes, and the way work flows through the organization. This was perhaps best described in the early 20th century by Henri Fayol as existing in the five functions of management: to plan, organize, coordinate, command, and control (Fayol 1984).

The traditional hierarchical organization structure (Figure 7.2) is represented by the traditional organization chart that we all know and love with the manager at the top and the many layers below. This is a typical command and control structure

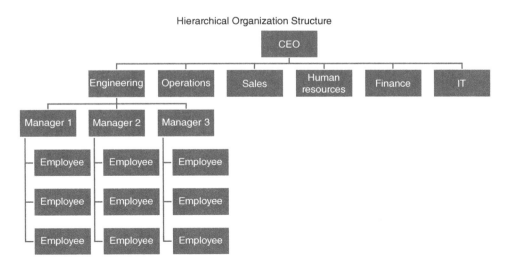

FIGURE 7.2 Hierarchical organization structure example.

(Fayol 1984) with a decision-maker at the top with a staff below them, who each have a staff below them, and so on. This is a centralized organization model. In this model, communication generally flows downward, not upward. This is how organizations were formed and managed since the late 19th century and in fact many of the fundamental theories that emerged during this industrial revolution were mechanical in nature based on this type of hierarchical structure. Yet today we understand that the organization is more than a machine and the person is more than the cog in that machine and new models have emerged.

The project-based organization (Figure 7.3) is one that is generally hierarchical in nature but operates entirely around specific and well-defined set of projects. A project is a ". . . temporary endeavor undertaken to create a unique product or service" (Project Management Institute 2000). In other words, this structure forms semi-permanent teams that are assigned to a single project from concept to conclusion. This offers the benefit of being able to focus the teams' efforts on a single product but suffers from not being able to see what is happening in other projects or parts of the organization. It also has the benefit of being able to dissolve the team at the end of the project. One downfall of the pure project organization is that the team members may not be highly incentivized to close out the project, as they will effectively be out of a job. In most cases, team members are reassigned to new projects, but in organizations that use contract workers, team members may need to find their own next project. If not actively managed, the projects can become internal islands that separate and divide the organization and live on long after they should have been dissolved.

Somewhere in between the hierarchical, the open system, and the project structure, there is the matrix organization structure (Figure 7.4). In the matrix organization, there are functional departments, which may include engineering functions such as systems, electronics, electrical, mechanical, thermal, test, electrical, safety, and computer-aided design (CAD). These resources are assigned to support multiple projects across the organization, hence the matrix. The matrix structure is also a project structure, so engineers from across the various functions may apply their efforts to a common project while having a functional reporting structure. In this manner, engineers can apply their skills across multiple projects that are in different stages of development. Which means that the total number of people in an organization may be reduced as people are supporting more than one project. There are variations within the project- or matrix-type organization. Table 7.1 compares the authority of the project manager, the role of the project manager (PM), and the amount of time functional groups team members are spending on project work. In the weak matrix structure, both the project manager and the team members only dedicate a small amount of their time to the project. An example of this could be the execution of an internal project to plan an event. The team may consist of members from management, marketing, finance, human resources, IT, and other functions who spend just a few hours a week on the project in addition to their regular functional duties. At the other end of the matrix spectrum is the strong matrix organization. In this structure, the project manager and team members are nearly fully dedicated to the project. This was the case in many of the large vehicle auto manufacturers who would set up a new project team to develop and launch a new vehicle line. The project manager in this instance is dedicated to this project and most of the team members will spend most of their time working on it.

There are a wide variety of other organizational structures that may be applied when designing your engineering team, and you may find that you utilize parts of several of them. In this model, the organization is viewed as ". . . an organism seeking

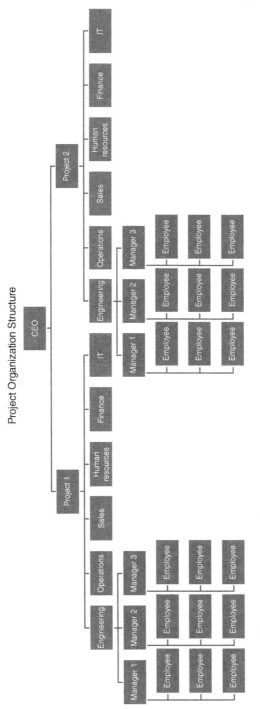

FIGURE 7.3 Project-based organization structure example.

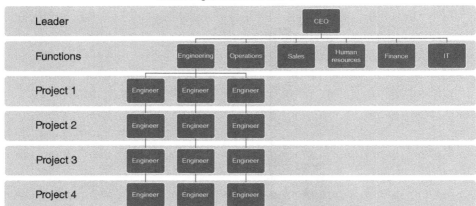

FIGURE 7.4 Matrix organization structure example.

TABLE 7.1 Types of Project Organizations

	Hierarchical	Weak Matrix	Balanced Matrix	Strong Matrix	Project
Project Manager Authority	None	Limited	Low to moderate	Moderate to high	High to total
PM Role	Part-time	Part-time	Full-time	Full-time	Full-time
% Time Team Spends on Project Work	Little	0–25%	15–60%	50–95%	85–100%

(modified from Project Management Institute PMBOK 2000).

to adapt and survive in a changing environment" (Morgan 1998, p. 35). The idea with this is that organizations may be considered as clusters of interconnected humans, businesses, and technologies. These clusters are affected by inputs from the outside such as the regulatory, financial, markets, constituents, stakeholders, and the geopolitical environments. In this model, the organization grows in a natural manner; in other words, it may be less structured, to make it more adaptable to market changes. The organization may have silos, like organs, that operate together but independently.

Late in the 20th century, the idea of the learning organization emerged. In this model, the organization is like that of an open system, but it applies a feedback loop and distributes knowledge throughout the organization. The learning model is often referred to as imagining organizations as human brains. "As we move into a knowledge-based economy where information, knowledge, and learning are key resources, the inspiration of a living, learning brain provides a powerful image for creation organizations ideally suited to the requirements of a digital age" (Morgan 1998, p. 69).

An organizational model was introduced by Dr. John Warner in 2003 and referred to as post-heroic leadership (2003). The concept of this doctoral research was that as the world has shifted into an information and knowledge economy, the organizations have evolved also into team-based organizations. Modern organizations are hierarchical and are based on having that "heroic leader" at the top of the organization, but they are leading a team of leaders. Each of those functional leaders is in turn leading a team of leaders. In the post-heroic organization, the leader is also leading horizontally and upwards and becomes more of an influencer in other areas of the organization. So instead of leading a team of followers, you are leading a team of leaders which changes the discussion. Post-heroic leadership includes beliefs and actions like having a shared sense of responsibility, a common vision, having mutually influencing relationships, has a focus on continuous development, is open to risk-taking, uses collaboration, and enables others to act while working toward the interests of the group (Warner 2003). This idea was demonstrated in the book *"Leadership Lessons of the Navy Seals"* (Cannon and Cannon 2003). The U.S. Navy Seals may be a textbook example of a post-heroic team. Each one of them is a leader within their team and each one is a specialist in a certain area. They have clear objectives and use flexible and dynamic organizational structures to accomplish those missions. They do this through developing trusting and loyal relationships.

PRACTICAL INSIGHTS | Generative Engineering

Engineering team structures evolve with an organization's business needs. They also evolve with technological advances. Throughout the early 2020s, the rapid development of artificial intelligence and machine learning led to advances in simulation tools for engineering teams. One type of simulation is **generative design.** *In 2024, SAE International published a research report in its EDGE series profiling this design practice and its impact on engineering teams, particularly those tasked with designing vehicle structures. This excerpt defines generative design and provides a guideline for how engineering teams, particularly those tasked with new vehicles such as EVs, can use it. Generative design can help optimize structures for weight, a particular concern for EVs due to their batteries.*

Generative design means different things to different people. It is used to describe various semi-automated methods for the design of physical structures using techniques including physics-based simulation, optimization, and machine learning (ML). Although generative design is most closely associated with structural optimization, the objective of the design optimization may include and combine properties such as stress, weight, manufacturability, heat transfer, fluid flow, or any other physical properties that can be simulated. By using ML, it may also be possible to include considerations of aesthetics within the objective.

Some see generative design as simply a topology optimization with multiple objectives – that are not combined into a single weighted objective function – which produces multiple possible designs from which the user can choose. Others see it as requiring less initial human definition than topology optimization, although this definition is disputed. In SAE's EDGE Research report, the term is used broadly to describe automated design methods, which can use various simulation, size, shape, and topology optimization; field optimization; and ML methods to produce designs optimized for multiple criteria including performance and cost. This may also be generically referred to as "computational design."

Generative design is not the same as generative AI, although there is some overlap between them. Generative AI uses a deep learning model to identify patterns in large sets of training data and can then generate equivalent new data, often from a simple text

(continued)

> **PRACTICAL INSIGHTS | Generative Engineering** (*continued*)
>
> prompt. Widely used examples include large language models (LLMs) (e.g. ChatGPT) and text-to-image systems (e.g. Midjourney). Such an approach could be used to reduce the requirement for physics-based simulation but is not currently in widespread use.
>
> Artificial neural network (ANN)-based ML and deep learning are seen as having the potential to improve computational design in two distinct areas: streamlining the user interface for a conventional generative design that uses physics-based simulations and reducing or removing the need for physics-based simulations altogether. LLMs such as ChatGPT could be used to streamline the user interface, allowing an engineer with little knowledge of simulation software to interact with a simulation model in much the same way they might currently interact with a simulation engineer. In the latter case, an AI model is trained using a number of designs, which have known performance as a result of previous simulation or testing. The model learns to recognize patterns and can then predict how new designs are likely to perform, without requiring explicit physics-based simulation, in a process akin to the intuitive understanding of designs that a designer gains through experience.
>
> *Like artificial intelligence (AI), generative engineering is still evolving. It is also not free of controversy. There are those who believe there is no substitute for physics-based modeling. Others see it as a natural extension of design modeling.*
>
> Source: Muelaner, J., "Generative Design in Aerospace and Automotive Structures," SAE Research Report EPR2024016, 2024, **https://doi.org/10.4271/EPR2024016**.

In reviewing these sources, the message is clear, we cannot design an engineering team independent of the rest of the organization or the product. Von Bertalanffy (1968) talked about the open system and the organismic system. In the first case, the open system is one that is influenced by forces from outside that system. In the system case, the organismic system is one that is designed like a living organism. Which makes it an open system. The challenge for the modern engineering manager is to develop processes and structures that can be fluid when they need to be and rigid at other times. The battery industry is also in such a constant state of change and flux that having a flexible engineering structure may better allow you to remain aligned with these changing demands. And in doing this, they must integrate a variety of touch points with the rest of the organization to facilitate cross-company communications. One of the key goals of an engineering leader is ensuring that the different engineering teams and functions are fully integrated and in constant communication with each other.

It is also interesting to note that von Bertalanffy (1968) describes the fact that in scientific fields, it is common for two very different scientific areas of study to solve the same problem independently. Which means that you should not limit your potential field of solutions to those you are already familiar with, you may find new and interesting solutions in unique areas. You may even think of bringing in experts in entirely new fields to give you a different perspective on your product and processes.

Designing an Organization

One engineering executive I recently spoke with said "... the structure of your engineering organization defines the structure of your product." What this means is that the final design and performance of the product you are developing will be largely dependent on how you organize your engineering team around that product. If you have a software-based organization, it will be the software that will drive the rest of the product. It also means that if your structure is not well-integrated, then the end-product integration will also likely be poor. This belief is called Conway's Law, named after Melvin Conway. Conway (1968) wrote that "... organizations which design systems (used in the broad sense here) are constrained to produce designs which are copies of the communication structures of these organizations". He goes on to write that the very act of designing an organization means that certain design decisions have already been made and these will be exhibited in the final product. In his 1969 foundational work on general systems theory, Ludwig von Bertalanffy wrote "It is necessary to study not only the parts and processes in isolation, but also to solve the decisive problems found in the organization and order unifying them, resulting from the dynamic interaction of parts, and making the behavior of parts different when studied in isolation or within the whole" (p. 31). Aidan Feldman adds that if you don't, your organization will struggle anywhere the product crosses organizational boundaries (Feldman 2021).

Knowing the different types of organizations that can be used is a great start, but in fact most often we do not actively plan what our organization looks like when our organization is young. In fact, few organizations spend time thinking about organizational structure, in many cases, they are trying to solve problems and issues quickly. But as we look at organizations in different stages of development, we see that the structure of their organizations is largely related to the stage of development they are in. Some of the problems that can be experienced when building an engineering team depend on the stage the company is in.

Corporate Culture

There are two parts to designing an organization. The first is designing the hierarchy, command, and control structure. The second is somewhat more difficult to purposefully develop but can be even more important, designing the culture of the organization. Culture is frequently described as "the way we do things around here." Deal and Kenney (1982), who literally wrote the book on corporate culture in the 1980s, describe it as "... a system of informal rules that spells out how people are to behave most of the time" (p. 15). Ten years later, Kotter and Heskett (1992) built on this by describing corporate culture as having two parts. The first is what they refer to as the invisible level which is harder to change, this is the shared values of a group. These persist over time even after changes in membership of the group or team. The second is easier to change and more visible, which is the behavior patterns of the group. These are the group behaviors that new employees are expected to follow by their fellow employees.

In building the culture of a team or company, there are several elements that must be carefully considered, including the business environment that the team and company operate in. This is largely related to the stage of development of the company and the field or markets you operate in. Next are the corporate values, the basic beliefs of the team and group. They are often memorialized at the company level as the corporate values, vision, and mission. However, within the departments and functions of an organization, there are sub-values that are created and encouraged by the departmental and functional leaders. These are almost never documented but are demonstrated by the behaviors of the leaders and are replicated by the team leaders. The next aspect of culture is the heroes of the team or organization. Who do the team members look up to or idolize? Who do they try to emulate? Who are the role models? In young start-up type companies and early-stage research and development companies, it is often the founder or inventor of the technology, but as the company grows and matures this often changes also. The next aspect of corporate culture is the rites and rituals of the group and team. How do we memorialize the small and big wins? What do we celebrate? These are the ". . . systematic and programmed routines of day-to-day life . . ." that show our employees the kind of behaviors that are expected. The final aspect of culture is the cultural network. This is the information communication channels within the team or organization. You may know it as the grapevine or rumor mill of the group and company. The cultural network is the carrier of corporate values, heroes, stories, and mythologies (Deal and Kennedy 1982).

In a McKinsey article, Chakravarty et al. (2021) write that "Startup or enterprise, incumbent or attacker, organizations with great engineering capabilities create a culture that fosters innovation and excitement alongside responsibility and care." Organizational culture takes time and work to build and must be actively managed. Culture is often referred to as "the way we do things around here" but more accurately it is the culmination of processes, beliefs, symbols, heroes, and actions that the team takes. Chakravarty et al. (2021) go on to list 10 principles that can be used to build your organization. You will notice in their list some of the same topics that we have already discussed, like keeping the team small and promoting learning. Their list includes:

1. Recognize that engineering is a craft not a commodity.
2. The leader must promote constant learning.
3. Coaching should be the main role of the engineering leader.
4. Develop flexible progression programs to allow your team the opportunity to grow.
5. Incorporate responsible development, an example of which may be designing your products for the circular economy.
6. Commit to diverse development, the more diverse the team the better the engineered solution.
7. Understand that tech debt is not always bad and that sometimes rework is ok, understand the tradeoffs.
8. Keep teams small.
9. Pursue relentless automation.
10. Embrace "everything as code" (Chakravarty et al. 2021).

More recently in a Forbes article, Peterson (2022) writes that engineers need three things: (1) to be able to be innovative, (2) to see the impact of their work, and (3) to be recognized for doing a good job. As CTO of the AI company Dialpad, some of the things Peterson does to build the culture around these three things include events he calls "hackathons," where engineers come together to join teams to invent and work out of the box and then present their results in a Shark Tank-like manner at the end of the one-week session. He also describes using mentorships with new engineers by partnering them up with an experienced engineer who already knows the ropes. Finally, he describes the importance of being flexible. These may not be right for your engineering team, but they give you an idea of some very out of the box tools you can use to help build and maintain the culture of the team.

In a 2022 Engineering and Leadership Podcast hosted by Pat Sweet, Mark Kinsella, VP Engineering at OpenDoor describes culture as "... the internal values and branding of the engineering team. The day-to-day processes that define, guide and reinforce those values throughout the engineering work" (Sweet 2022). Kinsella goes on to describe the value of the culture of the engineering team as being the x-factor that both keeps engineers moving quickly and helps to attract high-quality talent in a highly competitive environment. He talks about a strong culture being one where "people want to work together with other smart people on big problems." Kinsella also promotes the use of the idea of creating a buddy system, especially for new employees, interns, and co-ops.

There may be many examples of successes and failures in the battery industry when it comes to company culture. On the success side, we may consider the company Boston-Power. While they are no longer in business, the company was built around the idea of building a strong culture. In fact, the founder and CEO Christina Lampe-Onnerud interviewed many candidates and had a consultant whose main job was to interview prospective candidates to ensure they fit with the culture. Through the years, the culture of the company grew and flourished even as their technology struggled to get adopted. Even today you may be at a battery conference and see several former Boston-Power employees meet like they are family. On the other hand, battery companies have emerged and later failed, some of which could be tied to the culture of the companies. Companies like Romeo Power, EnerDel, BritishVolt, and even A123 could be considered in this realm. All were darlings of the investment community for a time, but eventually failed.

Engineering Teams in Start-Up Companies

Designing an organizational structure for a start-up engineering team is often the easiest because you are very often the technical inventor or innovator so you may fill the role of CTO as well as CEO. The organization will be composed of a handful of very highly skilled engineers or scientists that are often in your own network. The younger the organization the more you will rely on fewer, more highly experienced engineers with multiple fields of expertise. There is often a core group of engineers or scientists who are instrumental in developing the initial concept or invention. These team members may have a specialty, but they are more likely to operate like systems engineers as they are working closely together on all aspects of the technology. This team takes responsibility for the overall concept, but they may not have some of the specialties needed in later phases. This team will likely evolve to become the engineering leaders of the organization as it moves into the next stage of development.

The communication flow within the engineering team is smooth because there are no significant organizational layers to deal with. However, there may still be a certain lack of clarity of expectations with the other functions. Part of the role of the engineering leader is to ensure communication between functions is clear and understood.

As the start-up grows and raises more capital, the engineering team also begins to grow, bringing on board engineering vice presidents (VPs) or directors who are very hands on systems thinkers. The team here is focused on getting the concept to a preliminary state. Initial requirements may not be well documented at this stage and the team is building those requirements as they go.

In the start-up phase, there are also product-aligned organizations and technology-aligned organizations. A technology-aligned organization is most frequently seen in early-stage start-ups that are focused on developing a unique technology that integrates into a larger system. In the battery industry, we may think of this as chemistry, materials development, or cell development. A product-based organization is one that is focused on delivering a final product or service such as a battery pack that integrates into a vehicle.

Engineering Teams in Expansion Companies

As the organization grows out of the start-up phase and moves into the expansion phase, it becomes necessary to begin putting more structure in place. At this stage, the engineering team begins putting processes and tools in place to begin automating the work. The organization begins to grow by bringing in specialists in certain areas. Yet despite this, the engineering leader is still striving to keep the organization small and to avoid adding unnecessary complexity to the team.

The other challenge the engineering leader faces during this stage is managing uncertainty. Organizations in this stage still face much uncertainty in the market as their first products are yet unproven and early investors are looking for an early exit. The organization is highly impacted and influenced by changes in the financial markets. A change in the market could cause the organization to need to look at pulling its expansion back a bit and rightsizing the organization to make it through the storm. Another type of change to be considered is regulatory change. One example of this may be seen in the European Union's 2023 modifications to the Battery Directive. If you were developing a battery during the time this was underdevelopment, you may find that you need to make changes to your product now that it has been finalized and turned into law. The law is focused on ensuring that batteries have a low carbon footprint, use minimal harmful resources, use less non-EU materials, and are collected, reused, and recycled within the EU. According to the regulation, companies must identify, prevent, and address social and environmental risks linked to raw materials. It includes performance, durability, and safety requirements and requires new labeling, tracing, and tracking (Publications Office of the European Union, 2024). Rapidly changing regulations like this may require changes to product design or documentation to be able to operate in this region. Part of the difficulty with this is that these regulations are somewhat unique in that they do not exist in other regions. It can be difficult to make accommodations to international and global regulatory bodies. Because of this, it is important to maintain a small and flexible core team. You may bring in contract resources to meet some of your specialized needs. The main message at this stage may be to remain flexible.

PRACTICAL INSIGHTS | Li-Bridge Workforce Initiatives

As a result of the work done by the U.S. Department of Energy's Vehicle Technologies Office (DOE VTO) and the publication of their "National Blueprint for Lithium Batteries" in 2021, the organization Li-Bridge was created. Li-Bridge is a public–private partnership alliance that is focused on accelerating the development of a robust and secure domestic supply chain for LIBs. It is managed by Argonne National Labs and brings together industry trade groups NAATBatt International, New York Battery and Energy

> Storage Technology Consortium (NY-Best), and New Energy Nexus with the Department of Energy and other government entities.
>
> From this, Li-Bridge built on the National Blueprint included five goals, including number 5 "Maintain and advance U.S. battery technology leadership by strongly supporting scientific R&D, STEM education, and workforce development." This translated into Challenge #6 "Lack of domestic technical know-how, especially in mid-stream activities." This was identified as one of the greatest challenges to expanding domestic battery manufacturing in the United States. While the United States has a strong manufacturing base, it has not invested in educating and building the battery manufacturing base to meet the needs of the 2030 market. Jobs in the fields of battery-grade material processing, active material and component production, and end-of-life battery logistics were all identified as high need areas where there have been few current resources. As we move into the second half of the 2020s, there will be a huge demand for skilled workers in all areas of the battery space and organizations like SAE International, among others, are working to create curricula and programs to help educate and transition workers to fit these needs.

This stage also marks the point at which your first products begin to reach the end customers, which often requires the support of someone in an application engineering role who can support the customers' integration and return to the organization with recommendations to improve and bugs to work out.

Engineering Teams in Growth Companies

The larger the organization, the more complex the structure. This is true for organizations that have entered the stage of a growth company. The engineering team at this stage becomes more complex with its own hierarchy and its own substructure within the company. As the company begins developing new products and entering new markets, the engineering team gets somewhat more specialized with fewer engineers that have a good perspective on the whole.

Kinsella (Sweet 2022) said that as an engineering organization grows from 100 people, to 200, 300, 500, or more will drive a need to focus more on different areas to keep everyone working together and collaborating by focusing on communicating well. As the organization grows, it may be important to move from synchronous communication to a more asynchronous type of communication model. This enables remote work, global engineering organizations, and more well-thought on our communications models.

Engineering Teams in Mature/Declining Companies

As an organization reaches maturity and decline, it begins facing new challenges. In a mature organization, complexity reaches its greatest levels, communication becomes challenging, and subgroups tend to begin forming their own cultures and values. The engineering team is mainly concerned with managing a current portfolio of products and any new, innovative projects tend to be spun off into separate organizations to avoid impacting the current portfolio products. This tends to create an organization where there may be stagnation. It may see new, young employees join only to leave a few years later as they look for more challenging and rewarding opportunities.

During the decline stage of a company, the market acceptance of the products begins to decline as new and more innovative technologies replace them. At this point, one of two things happens, either new or innovative products are developed and the company moves into a renewal stage where it moves back one or two stages or it moves into the decline stage where it will ultimately die.

Challenges Associated with This Topic

Competition for people is one of the greatest challenges the industry faces today. With the growing demand for people across the entire value chain, companies are competing for the same group of employees. In the United States, there is a "battery belt" forming with many companies focusing on installing capacity in the same areas as shown in Figure 7.5. This image shows the announced and operational cell manufacturing plants which are clearly centered in the mid-west and southern regions. Battery pack manufacturing is not overlaid on this map but effectively doubles the number of facilities in the same regions. You will notice that there are four cell gigafactories in Tennessee and two more close by in Kentucky (Plante and Rindels 2022). This will create significant opportunities for job seekers and give them a chance to get competing offers. But from a hiring company perspective, this makes jobs more expensive and more difficult to fill.

This issue is exacerbated if you are trying to find people with direct experience in the battery industry, especially if you are looking for experienced battery engineers. McKinsey writes that "Attracting top talent is critical to achieving cost leadership" (Campagnol et al. 2022). They go on to note that this may be difficult to do because many of the major industry players have scooped up many of the available global industry experts in the field. With this much competition, companies will work hard

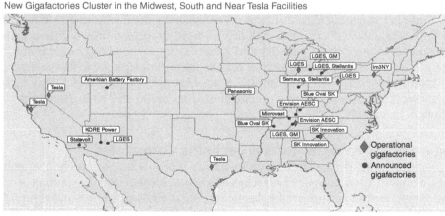

NOTES: We define a gigafactory as having capacity of 1 gigawatt hour or greater. Map does not include announcements missing a location or capacity estimate.

FIGURE 7.5 U.S. battery belt.
Source: Courtesy of Federal Reserve Bank of Dallas.

to retain their employees through retention bonuses, higher salaries, and extremely competitive benefit packages.

At the same time, universities have been slow to respond to the demand for degreed battery engineers, continuing to focus on advanced PhD and master's level programs (Campagnol et al. 2022). This has left a gap in the industry with newly minted engineering graduates emerging without good working knowledge of battery engineering. Most university programs offer classes in different aspects of battery engineering, but they do not offer complete battery engineering degree programs. Many universities are working on reconfiguring some of their existing programs to provide better battery education but there are many gaps. Part of the challenge for universities is that they already offer distinct mechanical, thermal, electrical, electronic, software, or chemical engineering programs. What they struggle with is that battery engineering requires a knowledge of all of these and more.

One clear trend that is occurring is that we are seeing a transition of engineers that are in the middle of their career from more traditional mechanical and electric components to the battery industry. In 2021, the Federal Consortium for Advanced Batteries (2021) (FCAB) published the *"National Blueprint for Lithium Batteries."* This included five goals, one of which was to "Maintain and advance U.S. battery technology leadership by strongly supporting scientific R&D, STEM education, and workforce development." Work in this area has been ongoing by organizations like NAATBatt International, whose battery education workgroup has provided many of the inputs into the FCAB blueprint and continues to work toward this goal of bringing new workers into the battery industry and training them. They work with groups like SAE International, who through a partnership with InnoBat are creating a set of training programs to help educate the engineering community. The electrochemical society (ECS) is also working on programs to help educate the scientific community. These are just a few examples of programs working to help retrain workers and give them the skills they need to be able to quickly adapt to the always changing battery industry. While not providing immediate help, these resources are actively working to help the engineer transition to become a battery engineer.

A challenge with all emerging and rapidly growing industries is managing the expansion and contraction cycles that naturally occur in that industry. Business fluctuations as experienced by industry expansion and contraction cycles can be caused by a variety of factors and may be categorized as being secular, seasonal, or cyclical trends, or miscellaneous random fluctuations. Secular trends are the natural upward and downward course of sustained development exhibited over long periods of time. Seasonal trends are those affected by the calendar year. Cyclical fluctuations are those activities that are characterized by recurring phases of expansion and contraction over relatively long periods. Miscellaneous random fluctuations are irregular, uncyclical fluctuations affecting businesses (Estey 1950). COVID-19 in 2020 and the Russia–Ukraine war in 2022 were perfect examples of a miscellaneous, random fluctuation as they could not be predicted but that had major impacts on the economy and growth of the industry. COVID-19 caused a near complete breakdown of the supply chain that took several years to recover from. The Russia–Ukraine war caused disruption to minerals coming from Russia such as nickel. A more secular fluctuation that conflicted with these was seen in the demand for new EVs and their batteries in the early 2020s which created greater demand than availability. This translated to an undersupply of cells and subsequently to the raw materials going into the cells which caused cell prices to skyrocket in 2021 and 2022.

These types of market fluctuations can be dangerous to the growth of an engineering organization if you do not manage that growth appropriately. Rapid expansion means filling unfilled jobs, contraction often means reducing headcount. The question becomes, how does a young and growing engineering team manage those growth cycles without causing permanent damage to the organization? How do you ensure that you can continue your development programs and keep them on track without losing key or significant engineering resources? The best method to manage this goes back to proper organizational planning. During the annual budgeting process, your engineering leadership should assess the programs that they are committed to executing along with an expectation of what new projects they may need to commit resources to. This activity when done along with some scenario planning will give you an idea what the minimum amount of resources may be needed, and the scenario planning will allow you to understand how you can juggle and manage those resources to accommodate more or fewer projects. In other words, overcoming these challenges may require developing accurate project estimates, prioritizing project activities, and optimizing resource allocation (Htet et al. 2023).

One method to help manage this is to utilize some contract engineering resources to fill some of your roles. Optimally, these may be contract-to-hire type resources that give you the opportunity to hire if they perform or if the role becomes permanent. This type of resource also allows you to quickly add or eliminate specific jobs quickly without damaging your "core" engineering team. Another method can be found in developing strategic partnerships. These can come in many forms. On the simplest level, you may decide to partner with an engineering services firm that can provide engineers with specific skill sets on a temporary basis for a short period. This allows the team to grow at a moderate level but with the possibility of filling some of the roles before they are needed on a full-time basis. Another strategic partnership may come from a strategic investor, who is also often a customer. This type of partnership may assign temporary engineering resources to a project that supports the development of their project. This brings forward the question of insourcing versus outsourcing your engineering projects or asking yourself should I develop it myself or can I work with an outside organization to develop it. In the younger organizations, it often makes sense to outsource some component groups. The battery management system is a good example of a subsystem that may be outsourced but this also leaves a knowledge gap in your organization.

Another resource that can provide support is the use of interns and co-ops students. Internships and co-ops provide valuable learning opportunities for the students especially if they are assigned to "real-world" projects and processes. While they may not have the industry experience, interns and co-ops can offer valuable support for the more senior engineers or program management, testing, and development activities. Typically, the students should be given a specific term project to work on with a defined outcome and a mentor to help them as they encounter challenges.

Case Study: American Battery Solutions

As introduced earlier, ABS is an example of a company that has passed through the start-up stage and moved into the expansion stage of development. The idea for the creation of "ABS" coalesced in mid-2018 and came to fruition in April 2019. The main idea of the company was to serve what was observed to be a segment of the market that was not getting much attention from the largest players – the lower-volume applications like delivery trucks and buses. These markets needed the same high-quality engineered products as the automotive segment but tended to struggle with access to cells and pack designs. ABS was founded to serve these customers. Much like most start-ups, the company was founded by an industry veteran who pulled together a small team of experts in various fields.

During this start-up stage, the company was founded by a very forward-looking family venture fund that enabled the acquisition of two facilities and funded the initial company and product development. To meet the company's goals of serving these markets, it was necessary to center the organization around two main areas: engineering and operations. The company was kept small, with a mandate to keep the headcount at a certain level. This meant the company worked exactly as described above with a small and highly experienced group of engineers who acted much like system engineers as they had responsibilities for multiple component subsystems. Within the engineering team, communication was relatively clear due to the small size of the team. However, one challenge that emerged quickly was also due to the small size of the teams. Product requirements were not well documented and project priorities were not always clear. At this stage of the company, an engineer may be supporting multiple projects and products rather than focusing their efforts on just one project.

This brought us to what I will define as the first crisis, the onset of the COVID-19 epidemic and the shutting down of nearly all businesses worldwide. Like nearly all companies, ABS moved to a work from home policy during the early months. With such a small team, this led to a variety of challenges including communication, IT, health of the team members, and project execution challenges. The team remained small during this period and as the stay in place orders began to be lifted by mid-2020 the team began growing. With several new project wins and a need to catch up on projects that had lost some momentum during COVID-19, the engineering team began growing by bringing on new engineers who had become available on the market as their prior companies downsized during COVID-19. While the engineering team began growing, there was still a high reliance on a few key team members due to their experience. Engineering processes and procedures were still in their infancy.

During the next few years, the company continued to grow at a slow but reasonable rate, doubling in size and then doubling again. This was the beginning of the expansion phase of the company when the first major customer projects were awarded, and the company began growing to support these projects. The engineering team expanded rapidly during this period, but the growth was somewhat variable. Because of the high demand for engineers in the battery field, the relatively young age of the company, and the high stress of trying to juggle multiple projects the team experienced some amount of regular turnover which kept the team from reaching their target headcount.

This led to what I will call the second crisis, which came in late 2022. While the team grew during this period, there was concern about the speed of the growth while the company was trying to manage a specific headcount target. The investors' concern over the speed of growth compared to the lack of revenue generation of a young company led to a need to "right-size" the organization.

(continued)

Case Study: American Battery Solutions (*Continued*)

This activity resulted in eliminating a portion of the overall headcount, the engineering team was hit particularly hard as it was the largest group in the organization. This led to a period of fluctuations as the engineering team restructured and worked to continue to support the existing programs. At this point, a very seasoned engineering leader was brought in to help restructure, grow, and operationalize the engineering team. He in turn brought in a team of leaders to help him manage the growing engineering team. The engineers began to become more differentiated in their roles and responsibilities and the team began developing more detailed processes and procedures, job descriptions, and responsibilities. The first products were developed and began being integrated into customer applications while the production lines were installed and operationalized.

At the same time, and as would be expected during the expansion stage, the company began evaluating different options to bring new capital into the company to support the expansion of the company. After several inquiries from several different groups, the original investors and the company were fortunate enough to conclude acquisition discussions with world-leading construction and mining OEM Komatsu Ltd. in December of 2023. This may indicate that the company is on the verge of moving out of the expansion phase and into the growth phase. As a wholly owned subsidiary of Komatsu, it may expect to have greater structure, governance, and oversight than it had during the first years of its existence.

This case study walks the reader through a company that has been through the first two stages of development and is beginning to move into the third stage. Literature has shown that all companies pass through a series of crises during their corporate life cycle but that the successful exit of each crisis leads the company to enter the next stage of its development.

Engineering Practices to Follow

Processes, the engineering design method is built around developing and executing processes. But a question that we should ask is do we have too many or too few processes? In the early or start-up company, there are not going to be very many processes in place. The team largely makes it up as they go along. But as the organization grows, it becomes important to be able to add processes and structure that support the needs of the growing organization.

Warner (2024), in Chapters 5 and 6 of "*The Handbook of Lithium-Ion Battery Pack Design Second Edition,*" outlined some of the processes associated with an engineering project including developing and executing a stage gate (or phase gate) process which can be used to bring a development program from concept to production. A properly managed stage gate process can be invaluable in ensuring alignment of the project with the organization as well as alignment with the original requirements. It is also a time to evaluate the results and status of the project. Many young organizations make the mistake of passing a gate as open rather than closing the gate to make corrections. For instance, perhaps the gate review identifies that the project is not long meeting the cost targets or is over budget or is not meeting the desired performance. This is the time to close the gate and give the team the time to resolve these issues. If the issues cannot be resolved, then the project is either rescoped or shut down. For example, I worked in the program management team for a large automaker on a new vehicle program. The program was unique in that its main purpose was to improve the access to the third row of seats by adding a third set of doors to the vehicle. The technology development was working well, and the program was meeting cost and performance targets. But at a gate just 18 months prior to production, the entire program was cancelled. During the gate review, it became clear that another vehicle development program that was underway and scheduled to launch a year later solved the same problem in a different and lower-cost manner. So rather than continuing the development, the learnings were documented and the program was closed, saving the company millions of dollars.

Within the engineering team, many other processes are critical for managing the product development. One of which is requirements management. When starting a new project, it is important to clearly document with as much detail as possible the engineering requirements. This allows you to check your status during the development, to manage change control, and it defines what testing should be done. This is often exemplified in the traditional automotive engineering "V" model which depicts the engineering requirements on the left side of the V and the testing that is used to validate that the product meets those requirements on the right side of the V. Tools such as Doors may be used to manage the requirements, but in smaller organizations it is largely done in Excel.

Engineering processes are iterative in nature. In their simplest form, they can be represented by the plan, do, check, act (PDCA) methodology. Start-up-type companies often push to minimize the iteration in the engineering process as they are focused on getting a product into the market quickly. However, the result of this is often that things get missed – requirements are not documented, usage is not understood, unknowns are . . . . well, unknown and not accounted for in the schedule. Sometimes decisions that were made in the first iteration are found not to be the correct decision and need to be modified. Iteration allows for continual improvement (engineeringtechnology.org 2024).

Other tools that are critical to engineering process are change control and issue tracking. Once a design has been iterated, it must be frozen or released into the system. This allows purchasing to buy the component to build it for testing. But this also gives you a baseline for measuring against. Whether you are measuring modifications to the released design or customer-driven changes, having a baseline design is critical to be able to manage the scope of the project. Without managing this level of change, you may find that the scope of the project has increased greatly, which will impact cost and timing and so must be closely managed. The other item mentioned above is issue tracking. It is important to be able to document issues as they arise, assign owners, and document resolutions. In a start-up company, this may be done in an Excel format but as the company grows, it will quickly outgrow this solution. One interesting solution that I have seen adopted is the use of SmartSheet. It is a tool that allows you to pull together many aspects of a program or project including schedule, BOM, costs, and issues and visualize them in a dashboard-like tool.

There are a wide variety of other tools and processes that may be integrated into your engineering teams including design of experiments (DOEs), which are used to explain and describe any variation in a design. One example of this was when we selected the spacing between the cells for a particular battery design. Several iterations were developed on a small scale and thermal runaway was initiated in one cell to evaluate the impact on the other cells at various distances. This chapter will not go into each of the different tools that are integrated into the engineering process, but they include a variety of software packages, CAD, computer-aided engineering (CAE), computer-aided manufacturing (CAM), modeling, software-in-the loop (SIL), and hardware-in-the loop (HIL) tools.

The last item, but perhaps one of the most important, to discuss is the alignment of goals and objectives. In much the same way that it is important to keep the values and vision of the engineering team aligned with that of the company, it is important that the individual goals and objectives of each engineer are aligned with the goals and objectives of the team, the project, and the company. As the adage says, it's not enough to get everyone in the same boat, they also need to be rowing in the same direction.

> **Future Considerations**
>
> For the engineering leader, the major consideration for the future should be flexibility, adaptability, and continuous learning. Take these and apply them to your organization, each will be unique, but you may find that most organizations will follow the development paths that are described here. The battery engineering field will continue to be an ever-changing landscape with new technologies and new chemistries emerging frequently. For a traditional battery company, think about splitting research and development. Research is focused on long-term and future technologies, whereas development is the process of engineering products for production. Yet, even within a rapidly changing world, you can build an engineering team that is appropriate for the stage of your company. Evaluate your team members closely as well as the candidates you are interviewing to see if they have the flexibility and skills needed for your organization.
>
> - Determine where your organization's activity is within the life cycle of the product. This will help you structure your team and guide its practices. Are you involved in materials research and development, or in short-term delivery of a product to market?
> - Advances in ML and artificial intelligence are finding their ways rapidly into engineering practices. Simulations and generative engineering – in which AI is used to develop physical structures without necessarily requiring the level of physics understanding normally required of a designer – are developing rapidly. How much engineers should rely on these technologies is unsettled.

Words to Know

Some of the key terminology used in this chapter is defined below, but for a more detailed description of terminology related to EVs and EV batteries, refer to SAE International standards J1715/1 and J1715/2.

Culture A set of shared values, beliefs, and norms that shape how people behave and interact in an organization.

Generative Design A computational design practice that uses ML to simulate a design without requiring the user to create all of the physics-based aspects of the design.

Hierarchical Organization An organizational structure that is defined in terms of having a "top-down" command and control structure where every person in the organization is subordinate to someone else.

Matrix Organization A matrix organization is one in which individuals may report to more than one leader across different groups or projects within an organization.

Post-heroic Leadership An organizational model in which a leader organizes but does not direct other leaders around a common vision. This approach tends to be less hierarchical and encourages risk-taking and collaboration. It constitutes a "two-way" traffic model of exchange and influence vs. a traditional command and control model.

Project A project is a temporary endeavor undertaken to create a unique product or service.

Project Organization A project organization is one in which the individuals in the organization are assigned to a specific project throughout the life of that project and then are reassigned to new projects at its conclusion.

Seven-Stage Organizational Model A conceptual model for identifying the various stages a business or company undergoes and for locating it within that stage or stages. The model helps clarify which approaches to engineering design teams are suited to the stage. Stages include start-up, expansion, professionalization, consolidation, diversification, integration, decline, and revitalization.

Team A group of individuals who perform interdependent and complimentary work to achieve a common goal or mission.

Further Reading

Here are publications we have found useful in building a general understanding of sustainability and of its role in many aspects of mobility discussed in this chapter.

Warner, John T. (May 2024). *The Handbook of Lithium-Ion Battery Pack Design: Chemistry, Components, Types, and Terminology.* (Second Edition) Cambridge, MA: Elsevier.

References

von Bertalanffy, L. (1968). *General Systems Theory: Foundations, Development, Applications* (Revised Edition ed.). New York: George Braziller.

Campagnol, N., Pfeiffer, A., and Tryggestad, C. (2022). Capturing the battery value-chain opportunity. **https://www.mckinsey.com/industries/electric-power-and-natural-gas/our-insights/capturing-the-battery-value-chain-opportunity** (accessed 4 May 2024).

Cannon, J. and Cannon, J. (2003). *Leadership Lessons of the U.S. Navy SEALS: Battle-Tested Strategies for Creating Successful Organizations and Inspiring Extraordinary.* New York: McGraw-Hill.

Chakravarty, A., Kerr, D., and Magoc, N. (2021). 10 principles that build great engineering organizations. **https://medium.com/digital-mckinsey/10-principles-that-build-great-engineering-organizations-2905382f1cec** (accessed 19 May 2024).

Conway, M. E. (1968). How do committees invent? *Datamation* magazine (April 1968), pp. 28–31. **http://www.melconway.com/Home/Committees_Paper.html** (accessed 20 April 2024).

Deal, T.E. and Kennedy, A.A. (1982). *Corporate Cultures: The Rites and Rituals of Corporate Life*. Reading, MA: Addison-Wesley Publishing Company Inc.

engineeringtechnology.org (2024). The engineering design process. **https://engineeringtechnology.org/engineering-graphics/the-engineering-design-process** (accessed 2 June 2024).

Estey, J.A. (1950). *Business Cycles: Their Nature, Cause, and Control*, 2nde. New York: Prentice-Hall.

Fayol, H. (1984). *General and Industrial Management*. (Revised Edition) (ed. I. Gray). New York: IEEE Press.

Federal Consortium for Advanced Batteries (2021). *National Blueprint for Lithium Batteries*. Washinton, DC: Office of Energy Efficiency & Renewable Energy.

Feldman, A. (2021). Designing engineering organizations. **https://jacobian.org/2021/jan/5/designing-engineering-organizations** (accessed 20 April 2024).

Flamholtz, E.G. and Randle, Y. (2016). Chapter 3: Identifying and surviving the first four stages of organizational growth. In: *Growing Pains: Building Sustainably Successful Organizations*, 5the (ed. E.G. Flamholtz and Y. Randle), 47–70. New Jersey, United States: Wiley **https://doi.org/10.1002/9781119176466.ch3**

Furr, N. and Dyer, J. (2020). Lessons from Tesla's approach to innovation. **https://hbr.org/2020/02/lessons-from-teslas-approach-to-innovation** (accessed 12 April 2024).

G&A Partners (2021). From idea to maturity, the five stages of business growth. **https://www.gnapartners.com/resources/articles/stages-business-growth** (accessed 14 April 2024).

Htet, A., Liana, S.R., Aung, T., and Bhaumik, A. (2023). Engineering management: a comprehensive review of challenges, trends, and best practices. *Journal of Engineering Education and Pedagogy* 1 (1): 1–7. Retrieved **https://www.researchgate.net/publication/371856339_Engineering_Management_A_Comprehensive_Review_of_Challenges_Trends_and_Best_Practices**

Inc. (2020). Organizational life cycle. *Inc.* (21 July) Retrieved from **https://www.inc.com/encyclopedia/organizational-life-cycle.html**

Kotter, J.P. and Heskett, J.L. (1992). *Corporate Culture and Performance*. New York: The Free Press.

McGowan, E. (2022). What is a startup company, anyway? **https://www.startups.com/library/expert-advice/what-is-a-startup-company** (accessed 14 April 2024).

Morgan, G. (1998). *Images of Organization: The Executive Edition*. San Francisco: Berrettt-Koehler Publishers, Inc.

Peterson, B. (2022). How to build an engineering culture that any company would envy. *Forbes* (18 February) **https://www.forbes.com/sites/forbestechcouncil/2022/02/18/how-to-build-an-engineering-culture-that-any-company-would-envy/?sh=e39660a1361b** (accessed 1 June 2024).

Petrone, P. (2016). The 6 stages every organization goes through as it matures. *LinkedIn.com* **https://www.linkedin.com/business/learning/blog/learning-and-development/the-6-stages-every-organization-goes-through-as-it-matures**

Plante, M. D. and Rindels, J. (2022). Automakers' bold plans for electric vehicles spur U.S. battery boom. *Federal Reserve Bank of Dallas*: **https://www.dallasfed.org/research/economics/2022/1011** (accessed 4 May 2024).

Project Management Institute (2000). *Project Management Body of Knowledge* (2000 Edition ed.). Newton Square, PA: Project Management Institute.

Roesch, R. (2020). Shrinking camera market: what Fujifilm should do in 2021 & beyond. **https://fujixweekly.com/2020/07/29/shrinking-camera-market-what-fujifilm-should-do-in-2021-beyond** (accessed 14 April 2024).

Senge, P. (1994). *The Fifth Discipline.* New York: Doubleday.

Smith, M. (2021). Back where we started: the camera industry is again a bit-part player. **https://petapixel.com/2021/11/26/back-where-we-started-the-camera-industry-is-again-a-bit-part-player** (accessed 14 April 2024).

Sweet, P. (2022). How to build brilliant engineering teams through incredible culture. **https://www.engineeringandleadership.com/engineering-culture** (accessed 1 June 2024).

Warner, J.T. (2003). *Post-Heroic Leadership: Mid-level Work Teams in Manufacturing Organizations.* Phoenix: ProQuest.

Warner, J.T. (2024). *The Handbook of Lithium-ion Battery Pack Design*, 2nde. Cambridge, MA: Elsevier.

Job Listings

An ongoing challenge when building a team or growing your organization is finding the right people and finding enough people to fill all open roles. The battery industry is growing but still remains a very small industry with few experts outside of R&D activities. According to the 2023 U.S. Energy and Employment Report (USEER), the clean energy workforce in the United States grew by almost 300,000 jobs in 2022, a 3.8% increase from 2021. Of this new job creation, jobs that were related to zero-emissions vehicles grew by 21%, adding 38,000 jobs (US Department of Energy, 2023).

The Bureau of Labor Statistics reports that in the zero-emissions, or electric vehicle space, these jobs are expected to be centered in three areas: (1) the design and development of electric vehicle models and their batteries, (2) the production of batteries, and (3) the installation and maintenance of the charging infrastructure and networks (Colato & Ice, 2023). Within these three areas, several specific jobs are expected to experience significant growth. In the design and development area, software engineers, electrical engineers, electronics engineers, and chemical engineering jobs are needed. In the battery manufacturing arena, the main needs will be around electrical, electronics, and electromechanical assemblers, and miscellaneous assemblers and fabricator roles. In the charging network field, the major roles needed will fall into the categories of urban and regional planners, electricians, electrical power-line installers and repairers, and construction laborers (Colato & Ice, 2023).

As can be clearly seen here, the types of roles that will be needed in different areas are directly related to the type of work done in each field. For the design and development of EVs and their batteries, it is the engineering roles that are most highly needed.

- Vice President, Engineering
- Chief Engineer
- Director, Engineering
- Battery Pack Lead
- Systems Engineer
- Mechanical Engineer
- Electrical Engineer
- Electronics Engineer, BMS
- Software Engineer, BMS
- Hardware Engineer, BMS
- Thermal Engineer
- Change Management
- Program Manager
- Designer
- Design Engineer
- Validation Engineer
- Manufacturing Engineer

References

Colato, J. and Ice, L. (2023). Charging into the future: the transition to electric vehicles. *Beyond the Numbers* 12 (4): Retrieved May 4, 2024, from **https://www.bls.gov/opub/btn/volume-12/charging-into-the-future-the-transition-to-electric-vehicles.htm#_edn22**.

US Department of Energy (2023). DOE report finds clean energy jobs grew in every state in 2022. **https://www.energy.gov/articles/doe-report-finds-clean-energy-jobs-grew-every-state-2022** (accessed 4 May 2024).

Electric Vehicle Batteries: From Sourcing to Second Life and Recycling, First Edition.
Edited by Bob Galyen and Frank Menchaca.
© 2025 John Wiley & Sons, Inc. Published 2025 by John Wiley & Sons, Inc.

Index

A

AC *see* alternating current (AC)
active balancing, 74–75
active materials, 9–14, 29, 30, 74
adsorption, 23, 42, 137
aerodynamics, 96
aerogels, 79
aged cells, safety of, 167–168
air cooling, 76, 79
alloys, hydrometallurgy from, 133
alternating current (AC), 84, 88, 97, 98, 100, 101
alumina, 30, 42, 60
   adsorption, 42
aluminum, 14, 35
American Battery Solutions (ABS), 194, 210–212
amperes, 66
anode, 48, 57–58, 137
   carbon nanotubes and graphene, 58
   composite, 58
   graphite, 57
   lithium metal, 58
   lithium titanate, 57
   materials, 12–14, 18, 30
   research and development, 14–18
   silicon, 58
   voltage *vs.* capacity curves, 18f
aqueous shredding, 135
aramid fibers, 60
artificial intelligence, 48
artificial intelligence interface, 44
artificial neural network (ANN), 202
artisanal mining, 43
automotive batteries, 48, 62, 73
aviation, 86–87

B

balance of system (BOS), 169
Battery as a Service (BaaS), 121, 155
battery belt, 208, 208f
battery design, 53–81
   battery pack, 68–75, 69f
      battery management system, 69–71
      state of balance, 74–75
      state of charge, 71
      state of energy, 75
      state of function, 73
      state of health, 72–73
      state of power, 74

cells, 54f
   anode, 57–58
   cathode, 56–57, 56f
   charge profile, 66
   cycle life, 65
   cylindrical, 54–56
   discharge power, 65–66
   electrolytes, 58–60
   energy content, 62–63
   energy density, 63
   internal resistance, 64–65
   maximum continuous discharge, 66–67
   maximum pulse discharge, 67
   nominal capacity, 62
   operating temperature, 63–64
   operating voltage, 63
   performance characteristics, 62–68
   pouch, 55, 56
   prismatic, 55, 56
   regenerative pulse charge, 67–68
   separator, 60–62
   temperature rise of continuous discharge, 68
cell to pack, 76–77
modular, 75–76
and structure, 75–77
thermal propagation, 77–80
battery electric vehicle (BEV), 36, 70–71, 86, 88–89
   and data privacy, 70–71
   powertrain, 88f
   propulsion system on motor vehicles, 94t
   typical sources for energy loss, 93f
battery-embedded energy, 93
battery energy storage systems (BESS), 97, 152, 159, 165, 168–169, 178
   failure, 169, 169f
   large-scale, 159
   root cause of failures, 169
battery management system (BMS), 38, 53, 69–71, 90, 139, 140, 157
battery pack, 68–75, 69f, 151
   battery management system, 69–71
   state of balance, 74–75
   state of charge, 71
   state of energy, 75
   state of function, 73
   state of health, 72–73
   state of power, 74
battery pack controller (BPC), 178

Electric Vehicle Batteries: From Sourcing to Second Life and Recycling, First Edition.
Edited by Bob Galyen and Frank Menchaca.
© 2025 John Wiley & Sons, Inc. Published 2025 by John Wiley & Sons, Inc.

battery pretreatments, 140–142
battery prices, limitation, 164
battery reuse network, 176
battery service, 145
battery's voltage measurement, 71
battery swapping, 119–122, 119t, 120f
beginning-of-life for application "B" (BoL$_B$), 155
BESS *see* battery energy storage systems (BESS)
BEV *see* battery electric vehicle (BEV)
bidirectional charging for electrified vehicles, 87
bidirectional power transfer (BPT), 116
bill of materials (BOMs), 129, 129t
binder-free electrode slurry, 144
Bipartisan Infrastructure Law, 113
black mass, 2, 132, 135
    fluoride contamination in, 144
blade battery, 77, 160
blockchain, 48
BMS *see* battery management system (BMS)
BOMs *see* bill of materials (BOMs)
brine, 1, 42
    resource, 33
    sources, 41
B2U Storage Solutions, 178

C

calendar aging, 155, 173
California Air Resources Board (CARB), 70, 170
California Consumer Privacy Act (CCPA), 70
CAM *see* cathode active material (CAM)
capacity fade, 72, 162
capital support, 45
carbon graphite, 12, 128
carbon intensity, 46
carbon nanotubes, 14, 58
carbon–sulfur composite, 22
carbon-sulfur composites with SPAN (CS-SPAN), 22, 23f
Catalyzing Innovative Research for Circular Use of Long-lived Advanced Rechargeables (CIRCULAR), 152
cathode, 56–57, 56f, 137, 153
    coprecipitation for synthesis, 134–135
    disordered rocksalt (DRX), 15–16, 15f, 16f
    electrode, 143
    LFP, 57
    manganese-rich NMC cathodes, 14, 15f
    material, 12, 30, 48
    NCA, 57
    NMC, 56, 57
    research and development, 14–18
cathode active material (CAM), 127, 128, 134
cathode healing™, 138
CCS *see* combined charging system (CCS)
cell chemistry, 56–68, 72, 76, 129–130
cell-to-cell thermal propagation, 85

cell-to-pack (CTP), 57, 68, 76–77, 160
centralized organization model, 198
ceramic-coated separators, 60
ceramics, 21, 79
certified energy (CE), 92
certified range (CR), 92
Charge de Move (CHAdeMO), 108
charge event reliability, 113
charge profile of cells, 66
ChargerHelp, 112–113
Charge Transfer Resistance, 64
Charging Interface Initiative e.V. (CharIN), 97, 102
charging power, 111
charging reliability of performance, 112
chemical compatibility, 60
chief executive officer (CEO), 192, 205
chief technology officer (CTO), 192, 205
circular economy, 5
    for EV batteries in Australia, 1–2
Clayton Valley lithium, 32
cobalt, 12, 35, 43, 46
    alternatives, 17–18
    ESG, 47
    higher energy density, 12
    structural stability, 17
    thermal stability, 17
cobalt oxide cathode, 143
combined charging system (CCS), 103, 104f, 119
commercial fleets, 105
commercialization process, 42
competitive market analysis, 49
composite anodes, 58
composite separators, 60
computational design, 201
connected and autonomous vehicle (CAV), 96
Contemporary Amperex Technologies, 76
contract engineering resources, 210
contract length and leverage, 44–45
control area network (CAN) bus, 69, 70
control pilot (CP), 100, 108
control systems, 169
conventional vehicles, 83, 84
Conway's Law, 203
cooling systems, 68, 77, 79
copper mines, 32
    extraction, 32
    global production, 32
coprecipitation for cathode synthesis, 134–135
corporate culture, 203–205
coulomb counting, 71, 72
COVID-19, 209, 211
C-rate, 66, 67, 153, 159
critical minerals, 30
cryo-shredding, 135
crystalline silicon cells, 32
CTP *see* cell-to-pack (CTP)
culture, defined, 203, 204

current interrupt device (CID), 167
cycle aging, 155
cycle life, 10, 18, 21, 22, 29, 57, 164
　capacity vs., 23f
　defined, 65
　factors influencing, 65
cylindrical cells, 28, 29, 54–56, 79, 129

D

damaged/defective/recalled (DDR), 128
data-driven methods, 171, 173
data fusion techniques, 72
data privacy, 70–71
DC see direct current (DC)
DCFC see direct current fast charging (DCFC)
decarbonized energy system, 32
decomposition temperature, 78, 79
deep discharges, 65
deep learning, 202
deep-sea dredging, 44
deep sea harvesting, 43–44
deep-state-of-health (deepSOH), 155
　to rescue, 173, 174f
degradation, 155–157, 156f, 158
dendrites, 21, 57
　formation, 61f, 62
　growth in solid-state lithium cells, 62
dendrite-suppressing technology, 21
depth-of-discharge (DOD), 151
design of experiments (DOEs), 213
dialysis membranes, 32
DICP see dissolvable ionic crosslinked polymers (DICP)
diesel engines, 36
diesel-powered mining vehicles, 36
digital product passport (DPP), 177
direct current (DC), 84, 88, 97, 98, 101
direct current fast charging (DCFC), 84, 102–104, 102t, 103f, 103t, 104f, 118
Direct Lithium Extraction (DLE), 41, 41f
　alumina adsorption, 42
　ion exchange method, 42
　membrane filtration, 42
　solvent extraction, 42
　use of, 41
direct recycling, 127, 138–140
　advantages, 140
　electrochemical reintroduction of lithium, 139
　hydrothermal methods, 139–140
　ionic solution methods, 139–140
　solid-state synthesis, 139
discharge power of cells, 65–66
discharging batteries, 140
disordered rocksalt (DRX) cathodes, 15–16, 15f, 16f
　cycling stability of, 15, 15f

dissolvable ionic crosslinked polymers (DICP), 143–144
distributed energy resource (DER), 108, 109, 117
DLE see Direct Lithium Extraction (DLE)
dry cells, 127
dry shredding, 136
DWPT see dynamic wireless power transfer (DWPT)
dynamic wireless power transfer (DWPT), 97, 115

E

e-bike batteries, 86, 111, 119
echelon utilization, 176
e-flight, 86–87
electrical abuse, 167
electrical and electronic (EE) powernet, 96
electrical energy, 88, 114, 140
electrical energy storage device, 95
Electric Car Sales, 34f
electric current, 111
electric propulsion system, 89, 90
electric vehicle (EV) batteries
　circular economy, in Australia, 1–2
　lifecycle, phases, 2
electric vehicle fires in Florida, 86
electric vehicle supply equipment (EVSE), 84–88, 100–102, 100f, 103f, 106, 108, 112–113
　infrastructure, 84–88
　operations and maintenance (O&M) training, 113
electric vehicle types, 88–91
　battery electric vehicle, 88–89
　hybrid electric vehicle, 89–91, 89f, 90f
　plug-in HEV, 91
　range-extended EV, 91
electric vertical takeoff and landing (eVTOL), 87
electrification, 32, 96, 110, 114–122
electrochemical cycling, 16
electrochemical hydrometallurgy, 134–135
electrochemical impedance spectroscopy (EIS), 163
electrochemical methods, 64, 72
electrochemical properties, 67
electrochemical reactions, 59
electrochemical reintroduction of lithium, 139
electrochemical society (ECS), 209
electrodes, 14, 32, 64, 67, 73, 136, 139, 140
　binder-free slurry, 143
　cathode, 143
　LMR, 15f
　negative, 38, 58, 59
　positive, 38, 58, 59
　solid, 61
　typical Li-ion battery with, 10f

electrode state of health (eSOH), 155
electrolytes, 10, 19, 23, 58–60, 137
  content of LFP cells, 79
  extraction, 136
  gel, 59
  high-concentration, 20
  hybrid, 60
  liquid, 20, 23, 59
  polymer, 59
  solid, 19t, 20, 21, 59
  solvents, 85
  superionic conductors, 60
  types of, 59–60
  wettability, 60
end-of-life asset (EOLA), 143
end-of-life (EOL), 126, 128, 140–142
end-of-life for application "A" (EoL$_A$), 155
end-of-life liability (EOLL), 143
End of Vehicle Life (EoVL), 153
end of warranty (EoW), 150, 156
energy capacity, 62–63
energy content of cells, 62–63
energy density, 20, 55, 61, 76, 77
  of cells, 63
  gravimetric, 63, 69, 86
  volumetric, 63, 86
energy management system (EMS), 169
energy storage system (ESS), 55, 111, 125, 157, 159, 166, 177
energy storage units, 166
energy transfer, 3
engineering teams and practices, 189–215
  battery belt, 208, 208f
  challenges, 208–212
    contract engineering resources, 210
    interns and co-ops, 210
    proper organizational planning, 210
    strategic partnership, 210
  designing new engineering organizations, 190–208
    corporate culture, 203–205
    expansion companies, 206–207
    growth companies, 207
    mature/declining companies, 207–208
    organism-based structure, 190
    organizational development stages, 191–197, 191f
    start-up companies, 205–206
    team-based structure, 190
    types of organizations, 197–202
  essentials of, 190
  expansion and contraction cycles, 208, 208f
  requirements management, 213
  stage gate process, 212
  stage of growth, 191f
entropy, 17
Environmental Protection Agency (EPA), 70

environmental social governance (ESG), 144
  in battery supply chain, 46–48
EOL see end-of-life (EOL)
EoW see end of warranty (EoW)
ESG see environmental social governance (ESG)
ESS see energy storage system (ESS)
ethyl carbonate, 31
evaporative methods, 32
EverBatt, 154f
EVSE see electric vehicle supply equipment (EVSE)
expansion companies, engineering teams in, 206–207
exponential growth, 30
extended producer responsibility (EPR), 177
external energy, 131
external supply chain, 46
extraction methods, 27, 36, 48, 132 see also Direct Lithium Extraction (DLE)
  electrolyte, 136
  evaporative methods, 32
  minerals, 2, 35

F
fast-charging batteries, 37
fire propagation mitigation, 55, 69, 76
fires, lithium-ion, 142
fire suppression system, 39
first life, battery, 162
fluoride contamination, 144
fluorophosphate decomposition products, 136
foreign entities of concern (FEOC), 48
formed cells, 128
fossil fuels, 3, 32, 36
fossil resource scarcity potential (FFP), 2
4M (man, machine, method, and material), 46
froth floatation, 137

G
Gaussian Process Regression, 163
gel electrolytes, 59
General Data Protection Regulation (GDPR) and Data Act, 70
generative design, 201–202
GHG emissions see greenhouse gas (GHG) emissions
global mining industry, considerations, 31–32
Global Technical Regulation (GTR), 170
global warming, 3
global warming potential (GWP), 2
granitic magmas, 33
graphene, 58
graphite, 13, 30, 33, 34, 48, 57, 106, 137, 138, 143
  natural, 57
  synthetic, 57
gravimetric energy density, 54, 63, 69, 86

greenhouse gas (GHG) emissions, 3, 46, 47
grid-forming protocol, 117
grid-storage demand, 158
grid-tied vehicle charging, 109t–110t
growth companies, engineering teams in, 207

H
hackathons, 204
hammer milling, 136
heat dissipation, 67
heat generation, 78, 79
heating, ventilation, and air-condition (HVAC), 101
heavy-duty vehicles, 105
HE-LMNO, 17
heroic leader, 201
heterogeneous aging, 74
HEV *see* hybrid electric vehicle (HEV)
High Capacity Battery Technology, 4, 9–24
high-concentration electrolytes, 20
high-frequency power line communications (PLCs) protocol, 108
high-voltage (HV), 84, 86, 95, 99
home energy battery pack (HEBP), 2
home energy storage system (HESS), 2
HPPC *see* hybrid pulse power characterization (HPPC)
hybrid electric vehicle (HEV), 67, 89–91, 89f, 90f
  parallel, 89f, 90
  series, 90, 90f
hybrid electrolytes, 60
hybrid pulse power characterization (HPPC), 163, 164f
hydrofluoric acid (HF), 95, 137
hydrogen peroxide decomposition, 133
hydrometallurgy, 131–135, 132f, 154
  from alloys, 133
  electrochemical, 134–135
  precipitated mixed metal oxides, 132
  purified metal salts, 132
  recycling from LIBs (without smelting), 133
  reprecipitation of cathodes from, 134–135, 135f, 138–145
    battery pretreatments, 140–142
    design and recycling, 142
    direct recycling, 139–140
    materials, 143–145
hydrophobicity, 137
hydroprocessing, 128
hydrothermal methods, 139
hygiene factors, 193

I
ICE *see* internal combustion engine (ICE)
ICEVs *see* internal combustion engine vehicles (ICEVs)
imagining organizations as human brains, 200
immersion cooling, 79
impedance spectroscopy, 72
industry standards, 166–167
Inflation Reduction Act (IRA), 3, 178
infrastructure, 83
  aerodynamics, 96
  assessing range requirements, 97–98, 98f
  availability of service, 112
  battery swapping, 119–122, 119t, 120f
  charge event reliability, 113
  design for battery clearance and deformation, 95–96
  designing for charging interoperability, 106–110, 107t
  drives battery and vehicle design, 92–122
  electric vehicle supply equipment, 84–88
  electrification movement affects OEM's technology roadmaps, 114–122
  new infrastructure affects supply chain, 110–114, 110f, 114t
  performance improvement, 111–112
  ride and handling, 96
  stranded energy, 95
  system safety, 98–99
  thermal management, 96–97
  types of charging, 99–122
    DC fast charging, 102–104, 102t, 103f, 103t, 104f
    local charging, 99–102, 99t, 100f, 101t
    low temperatures, 105–122
    vehicle types and considerations, 104–105
  weight management and its effects, 93–94, 94t
inorganic materials, 19t, 20
inorganic solid electrolytes, 59
instantaneous power, 74
intercalation, 23, 34
internal combustion engine (ICE), 83, 89–91, 114–122, 145
internal combustion engine vehicles (ICEVs), 3, 5
internal resistance, 67, 72, 105
  of cells, 64–65
    components, 64
    defined, 64
    effects, 64
    ohms, 64
    weak cells, 160
  state of health, 72
  state of power, 74
internal supply chain, 46
interoperability, 106–110, 107t, 112
In-vehicle Battery Durability for Electrified Vehicles, 92
in-vehicle infotainment (IVI), 96
ion exchange method, 42

226 Index

ion-exchange synthetic methods, 138
ionic conductivity, 59, 60
ionic solution methods, 139–140
iron phosphate cathode, 85
Irvine, 17
iteration strategies, 193

J

J2836 ecosystem, 109f

K

Kalman filter, 71
kinetic energy, 67
kinetic power, 92
knees, 162
knowledge-based economy, 200, 201

L

large language models (LLMs), 202
laterite, 43
layered oxide, 24
LCAs *see* lifecycle assessments (LCAs)
LCO *see* lithium cobalt oxide (LCO)
learning model, 200
legacy propulsion methods, 35
LESAs *see* lithium-electrode sub-assemblies (LESAs)
levelized cost of energy (LCOE), 165
LFP *see* lithium iron phosphate (LFP)
LHD *see* load, haul, and dump (LHD)
Li-Bridge workforce initiatives, 206–207
lifecycle assessments (LCAs), 2, 4, 47, 155
life cycle of EV battery, 151f, 153, 153f
Li-ion electrochemistry, 37
Li-metal anodes, 19–22, 20f
liquid catholyte, 19
liquid cooling, 76, 78, 79, 111, 169
liquid electrolytes, 20, 23, 59
liquid–liquid extraction (LLE), 133
lithium, 46
  applications, 33
  early uses, 33–35
  extraction, 32
  resources, 46
lithium-and manganese-rich (LMR) oxides, 14, 15f
lithium carbonate, 32, 42, 43
lithium carbonate equivalent (LCE), 32
lithium chloride, 32, 33
lithium cobalt oxide (LCO), 12, 34, 56
lithium-electrode sub-assemblies (LESAs), 21–22
lithium garnets (LLZO), 59
lithium hydroxide, 32, 43
lithium-intercalating oxides, 19
lithium-ion battery (LIB)
  active materials, 10–14, 11t
  anode materials, 12–13
  cathode materials, 12
  cells, 28–30, 29f
    applications, 30
    cylindrical cell, 29
    defined, 28
    electrolyte cell, 28f
    key components, 29
    pouch packaging, 29
    prismatic, 28
  commercialization of, 10, 33
  constituents of, 35
  with electrode reactions, 10f
  ESG in battery supply chain, 46–48
  evolution of, 34–35
  fast charging, 37
  Li-ion blend, 38
  lithium–sulfur and solid-state materials, 19–23
  opportunity charging, 37
  plug-and-play modules, 38–39
  recycling, 125–146
  sourcing, 27–52
    capital support, 45
    cell types and, 28–30, 29f
    challenges, 50–51
    cobalt, 43
    contract length and leverage, 44–45
    cost *vs.* price, 44–46
    deep sea harvesting, 43–44
    DLE, 41–42
    granitic magmas, 33
    hard rock sources, 42
    lithium supply, 40–43, 41f
    nickel, 43
    quality and, 45–46
    raw materials, 30–32
    regulations and standards for ethical, 50
    selection, 33–43
    subterranean brine reservoirs, 33
    supply chain complexity and challenges, 49
    supply chain management, 48–49
    technological innovations in, 48–49
  swapping, 37
  thermal propagation, 85
lithium-ion cells, 54f
  anode, 48, 57–58
  cathode, 56–57, 56f
  charge profile, 66
  cycle life, 65
  cylindrical, 54–56
  discharge power, 65–66
  electrolytes, 58–60
  energy content, 62–63
  energy density, 63
  internal resistance, 64–65

maximum continuous discharge, 66–67
maximum pulse discharge, 67
nominal capacity, 62
operating temperature, 63–64
operating voltage, 63
performance characteristics, 62–68
pouch, 55, 56
prismatic, 55, 56
regenerative pulse charge, 67–68
separator, 60–62
temperature rise of continuous discharge, 68
lithium iron phosphate (LFP), 12, 29, 32, 38, 56, 57, 78–80, 128, 131, 153–154, 174–175, 175f
lithium manganese batteries, 56
lithium manganese oxide (LMO), 38, 56
lithium metal anode, 58
lithium mine, 31f
lithium oxide, 32
lithium sulfate, 43
lithium titanate anodes, 57
lithium titanate oxide (LTO), 38, 137
    battery rate, 38
    drawback of, 38
lithium transition-metal oxide electrode materials, 138
LMO *see* lithium manganese oxide (LMO)
load, haul, and dump (LHD), 36
    scenarios, 39–40
local charging, 99–102, 99t, 100f, 101t
localization, 27, 49
loss of active material of the negative electrode (LAM$_{NE}$), 155, 156
loss of active material of the positive electrode (LAM$_{PE}$), 155, 156
loss of lithium inventory (LLI), 155
low temperatures, 105–122

M

machine learning (ML), 72, 201
magnesium salts, 42
management of chain (MOC), 46
manganese, 12, 35
manganese-rich NMC cathodes, 14, 15f
market evaluation price, 164
market fluctuations, 210
mass transport resistance, 64
material processing, 27, 48
matrix organization structure, 198, 200f
mature/declining companies, engineering teams in, 207–208
maximum continuous discharge, 66–67
maximum pulse discharge, 67
mechanical abuse, 167
mechanical stability, 58, 59
mechanical strength, 13, 14, 54, 58, 60

megawatt charging (MWC), 87, 118f
Megawatt Charging System (MCS), 118
membrane filtration, 42
metal casing, 54
metallic alloys, 131, 132
metallic lithium, 30, 62
metallurgy, 130
metal–organic complexes, 133
metal oxide, 48, 131, 132
    carbon separation from, 137
Methyl Carbonate, 30
minerals
    critical, 30
    extraction, 35
    hard rock, 46
minimal viable product approach, 193
mining
    battery electric vehicle, 36
    environmental impacts of, 31, 32
    lithium, 31f
    natural gas, 36
    vehicles, voltage of, 38
mobility devices, 110–111, 110f, 119
modularization of swappable batteries, 120f
modules, 75–76, 120
    selection and adequacy, 160
multi-material approaches, 94
MWC *see* megawatt charging (MWC)

N

NCA *see* nickel cobalt aluminum oxide (NCA)
net present value (NPV), 165
neutralization, 140, 142
new scrap recycling, 127–128
nickel, 12, 35, 43, 46, 57
    content cells, 54
    ESG, 46–47
    pricing, 128–129
    TM cathodes with changing, 12f
nickel cobalt aluminum oxide (NCA), 29, 32, 57
nickel, manganese, cobalt oxide (NMC), 29, 32, 38, 56, 57, 78, 128, 135, 138, 153
    high nickel content, 77, 129
    hydrometallurgy, 135, 138
    manganese-rich NMC cathodes, 14, 15f
    thermal stability characteristics
        decomposition temperature, 78
        heat generation, 78
        operating temperature range, 78
        thermal runaway, 78
Nio, Inc., 121
NMC *see* nickel, manganese, cobalt oxide (NMC)
NMC-811 cathodes, 12
n-methyl pyrrolidone (NMP), 144

nodules, 43, 44
nominal capacity of cell, 62
nominal voltage, 63

O

OBC *see* on-board charger (OBC)
OEMs *see* original equipment manufacturers (OEMs)
Ohmic Resistance, 64
old scrap recycling, 129
olivine, 12
on-board charger (OBC), 100–102, 106
on-board diagnostics (OBD), 70
Open Charge Point Interface (OCPI), 113
Open Charge Point Protocol (OCPP), 113
open system, 202
operating temperature of cells, 63–64
operating temperature range, 78, 79
operating voltage of cells, 63
opportunity charging, 37
Order for Zero Emission Vehicles (ZEV), 171
organic materials, 19*t*
organismic system, 202
organizational development stages, 191–197, 191*f*
  consolidation, 191, 191*f*
  decline and revitalization stages, 191, 191*f*, 196
  diversification, 191, 191*f*, 194
  expansion, 191, 193–195
  integration, 191, 191*f*, 194
  maturity stage, 195
  new venture creation, 192
  professionalization, 191, 191*f*
organizational structure, 190
  centralized organization model, 198
  defined, 197
  hierarchical, 197*f*
  learning model, 200
  matrix, 198
  project-based organization, 198
original equipment manufacturers (OEMs), 54, 88, 100, 110, 127, 195
  electrification movement affects, 114–122, 115*f*, 117*t*, 118*f*
  Nio, Inc., 121
oxygen cathodes, 22–23

P

parallel hybrid electric vehicle (PHEV), 89, 89*f*, 106
Paris Agreement, 3, 31
part submission warranties (PSW), 46
passenger vehicles, 105
passive balancing, 74
path dependency of degradation, 170, 173
pegmatites, 33

Personal Information Protection Law (PIPL), 70
personal mobility devices, 105
PEV *see* plug-in electric vehicle (PEV)
PFAS *see* polyfluorinated alkyl substances (PFAS)
phase change materials (PCMs), 78
phosphorus, 23, 35
physical methods, 42
plan, do, check, act (PDCA) methodology, 213
plug-and-charge, 108, 113, 114
  benefit of, 114
plug-in electric vehicle (PEV), 88, 96, 106
plug-in hybrid electric vehicle (PHEV), 91
polyacrylic acid (PAA), 143
polyacrylonitrile, 22
polyanions, 24
polyethylene, 85
polyethyleneimine (PEI), 143
polyethylene oxide (PEO), 20, 59
polyethylene (PE), 60
polyfluorinated alkyl substances (PFAS), 143, 145
polymer-based materials, 20
polymer-coated aluminum, 136
polymeric binders, 136
polymers, 20, 21, 29, 59, 60, 143
polymers of intrinsic microporosity (PIMs), 21, 22
polyolefins, 60
polypropylene (PP), 60, 85
porous polyethylene separators, 30
post-heroic leadership, 200, 201
pouch cells, 55, 56, 127, 129
power conversion systems (PCSs), 169
power density, 87
power fade, 72
power management strategies, 63, 77
precursor materials, 30
prismatic cells, 28, 55, 56, 77, 129
product-aligned organizations, 206
product coding system, 177
product development, 193
production part approval process (PPAP), 46
project-based organization structure, 198, 199*f*
  example, 199*f*
  types of, 200*t*
proof-of-concept, 159
propulsion system, 91*t*, 95, 99
protective earth (PE), 100
proximity pilot (PP), 100
Prussian blue (PB), 24
pulse wave modulated (PWM) signal, 108
pyrometallurgy, 130–132
pyroprocessing, 127, 128

Q
quality, and sourcing, 45–46
quality control, 166
quality management systems, 45

R
range-extended electric vehicle (REEV), 91
reagent, 41
   sulfuric acid, 42
recycling LIBs, 5, 125–146
   dry shredding, 136
   froth floatation, 137
   hammer milling, 136
   hydrometallurgy, 131–135, 132f
     from alloys, 133
     electrochemical, 134–135
     precipitated mixed metal oxides, 132
     purified metal salts, 132
     recycling from LIBs (without smelting), 133
     reprecipitation of cathodes from, 134–135, 135f, 138–145
   net battery recycling profit, 154f
   new scrap, 127–128
   old scrap, 128–129
   pretreatment and harvesting, 135
   pyrometallurgy, 130–132
   second-life before, 150–155
   technical methods, 129–137
   thermal treatment, 137
   value chain feedstock types, 127
ReElement Technologies, 155
reference performance test (RPT), 163
refurbishing heterogeneous battery packs, 157
regenerative braking systems, 67
regenerative charging, 90
regenerative pulse charge of cells, 67–68
ReJoule Energy, 171
reliability issues, 113, 114t
remaining useful life (RUL), 170, 173, 176
   estimation, 162–163
   prediction of, 173
remanufacturing, 157
renewable energy, 32
   sources of, 3, 36
   storage, 65
reprecipitation of cathodes from hydrometallurgical processing, 134–135, 135f, 138–145
RePurpose Energy, Inc., 171
resource-intensive process, 5
road load power, 92
RUL *see* remaining useful life (RUL)

S
SAEJ1939, 70
SAEJ3400, 107–108

safety hazards, 168–169, 169f
Saft's battery system, 37, 39
scrap, 48
second-life batteries (SLBs), 2, 5, 149
   battery reuse network, 176
   defined, 150
   diagnostics and prognostics, 170–175
     assessment, 170–174, 172f
     DeepSOH to rescue, 173, 174f
     LFP cells, 174–175, 175f
     path dependency of degradation, 173
   evaluation, 159
   implementation, 160–165
     economic viability and price limitations, 163–165
     extent of reutilization, 160–162, 161f
     module selection and adequacy, 160
     SOH and RUL estimation, 162–163, 163f
     state of market and technology, 162
   policies and financial incentives, 176–178
     echelon utilization, 176
     EPR, 177
     product coding system, 177
     traceability, 176, 177
   potential uses of, 158–159, 158t
   quasi-stationary, 158
   repurposing and, 150–157
     battery degradation, 155–157, 156f
     pathways to, 157
     before recycling, 150–155
     remanufacturing, 157
   safety, 165–169
     of aged cells, 167–168
     certification, 168, 168f
     hazards, 168–169, 169f
     industry standards, 166–167
   stationary, 158
second-life battery management system (SL BMS), 162
SEI *see* solid electrolyte interphase (SEI)
selective harvesting, 44
self-discharge rate, 72
sensor technology, 71
separator, 60–62, 137
   ceramic-coated, 60
   composite, 60
   improvement, 60
   properties for selection, 60
series hybrid electric vehicle, 90, 90f, 91
shallow discharges, 65
shredding batteries, 135–136
silicon, 30
   anodes, 18
   composite anodes, 13–14
   nanoparticles, 13
silicon-based anodes, 58
silver mines, 32

slag, 131
SLBs *see* second-life batteries (SLBs)
smelting, 130–132
SoC *see* state of charge (SoC)
sodium carbonate, 42
sodium-ion battery technology, 9, 23–24
sodium-ion electrochemistry, 58–59
sodium vanadium phosphate and fluoride (NVPF), 24
sodium vanadium phosphate (NVP), 24
SoH *see* state of health (SoH)
SOH-C *see* state of health–capacity (SOH-C)
SOH-R *see* state of health–resistance (SOH-R)
solid electrolyte interphase (SEI), 62, 85, 128, 139, 155, 172, 173
solid electrolytes, 19t, 20, 21, 59
solid-state batteries, 20, 61–62
solid-state energy storage technologies, 21
solid-state lithium metal batteries, 21
solid-state synthesis, 139
soluble binders, 144
solvent extraction, 42, 136
sorbents, 32
sourcing, LIBs, 27–52
  capital support, 45
  cell types and, 28–30, 29f
  challenges, 50–51
  cobalt, 43
  contract length and leverage, 44–45
  cost *vs.* price, 44–46
  deep sea harvesting, 43–44
  DLE, 41–42
  granitic magmas, 33
  hard rock sources, 42
  lithium supply, 40–43, 41f
  nickel, 43
  quality and, 45–46
  raw materials, 30–32
  regulations and standards for ethical, 50
  selection, 33–43
  subterranean brine reservoirs, 33
  supply chain complexity and challenges, 49
  supply chain management, 48–49
  technological innovations in, 48–49
spoke deactivation, 141–142
Standard Hydrogen Electrode (SHE), 24
start-up companies, engineering teams in, 205–206
start-up company, 213
state of balance (SoB), 74–75
state of certified energy (SOCE), 155
state of certified range (SOCR), 92, 155
state of charge (SoC), 38, 64, 66, 68, 71, 87, 135
state of energy (SoE), 75
state of function (SoF), 73
state of health–capacity (SOH-C), 155, 156, 170, 173

state of health–resistance (SOH-R), 155, 173
state of health (SoH), 38, 66, 72–73, 112, 153, 171–172
  aspects, 72
  challenges, 73
  determination method, 73
  estimation, 162–163, 163f, 171, 174
  measurement, 170
  techniques for, measure and estimate
    coulomb counting, 72
    electrochemical methods, 72
    impedance spectroscopy, 72
    machine learning models, 72
  testing methods, 171
  voltage *vs.*, 175f
state of market and technology, 162
state of power (SoP), 74
state of safety (SOS), 165
stranded energy, 95
strategic partnerships, 210
strong cells, 160
subterranean brine reservoirs, 33
sulfur, 20
  cathodes, 22
sulfuric acid, 42, 133
sulfurized polyacrylonitrile (SPAN), 22
superionic conductors, 60
supervisory control and data acquisition (SCADA), 178
supply chain, 27, 29, 31, 45
  bottlenecks, 49
  challenges, 49
  complexity, 49
  environmental social governance in, 46–48, 47f
  external, 46
  internal, 46
  new infrastructure affects, 110–114, 110f, 114t
  strategic planning for, 49
supply chain management, 48–49
supply–demand dynamics, 27
surplus ore potential (SOP), 2
sustainability, defined, 3
swappable batteries, 119–122, 119t, 120f

T

teardown model, 160
technoeconomic models, 23
technology-aligned organizations, 206
temperature rise of continuous discharge, 68
terrestrial acidification potential (TAP), 2
terrestrial-based mining processes, 44
Tesla, 54, 77, 107, 143, 195
text-to-image systems, 202
thermal abuse, 167
thermal conductivity, 79

thermal harvesting of harvest transition metals, 131
thermal interface materials, 77
thermal management systems, 39, 76, 96–97
thermal propagation, 77–80, 95
    anatomy of, 85
    improper charging, 78
    lithium iron phosphate, 78–80
    manufacturing defects, 78
    mitigation, 75
    NMC batteries, 78
    overcharging, 78
    physical damage, 78
thermal regulation, 75
thermal runaway, 64, 68, 78, 79, 84, 87, 95, 166
thermal stability, 60
    LFP batteries
        decomposition temperature, 79
        heat generation, 78
        operating temperature range, 78
        thermal runaway, 78
    NMC
        decomposition temperature, 78
        heat generation, 78
        operating temperature range, 78
        thermal runaway, 78
thermal treatment, 137
Tier 1 suppliers, 45
trade-offs, 5, 75
transparency, 48, 70
Twig Power, 192–193

U

UBE *see* usable battery energy (UBE)
uncycled cells, 167
unformed cells, 128
unused cells, 167
uranium, 33
urban mining, 36
usable battery energy (UBE), 91, 92, 150

V

vacuum drying, 136
value chain feedstock types, 127
vehicle demand energy (VDE), 92, 105
    equation, 92
    kinetic power, 92
    mitigation strategies, 93
    road load power, 92
vehicle mass, 93
Vehicle-to-Business (V2B), 117
vehicle-to-demand (V2X), 116, 118*f*
Vehicle-to-Grid (V2G), 117
Vehicle-to-Home (V2H), 117
Vehicle-to-Load (V2L), 116
Vehicle-to-Vehicle (V2V), 117
volumetric energy density, 55, 86

W

waste generation, 46
water consumption (WCP), 2
weak cells, 160
wet cells, 127
wet printing, 127
Whittingham, Stanley, 34–35
whole-pack model, 160
willing to sell price, 164
wireless charging network, 115*f*
    applications, 115
    benefits of, 115
Wireless Electric Road Systems (wERS), 115
wireless power transfer (WPT)
    levels, 116*t*
    operation of, 115
Worldwide Harmonized Light Vehicles Test Procedure (WLTP), 92
WPT *see* wireless power transfer (WPT)

Z

zero-emission vehicle (ZEV), 97